Normally-Off Computing

Takashi Nakada · Hiroshi Nakamura
Editors

Normally-Off Computing

Springer

Editors
Takashi Nakada
The University of Tokyo
Tokyo
Japan

Hiroshi Nakamura
The University of Tokyo
Tokyo
Japan

ISBN 978-4-431-56805-6 ISBN 978-4-431-56505-5 (eBook)
DOI 10.1007/978-4-431-56505-5

Softcover reprint of the hardcover 1st edition 2017

Printed on acid-free paper

This Springer imprint is published by Springer Nature
The registered company is Springer Japan KK
The registered company address is: Chiyoda First Bldg. East, 3-8-1 Nishi-Kanda, Chiyoda-ku, Tokyo 101-0065, Japan

Preface

Steady technology improvement enables us to realize rapidly evolving information society. Information devices help us anywhere and anytime in our daily life. In the society, these devices are expected to be as small as possible in size and keep their operations as long as possible for a wide variety of applications. Thus, improvement of energy efficiency is indispensable to realize long battery life with limited battery capacity.

This book is designed for both hardware and software designers. Energy consumption of modern information devices strongly depend on both of them. Software designers should not only understand the behavior of hardware but also optimize software to reduce power consumption. Hardware designers also have the same responsibility. Therefore, co-optimization of hardware and software is indispensable to improve energy efficiency.

This book is based on the outcome of the Normally-Off Computing project from 2011 to 2016 supported by NEDO/METI (New Energy and Industrial Technology Development Organization/Japanese Ministry of Economy, Trade and Industry). Normally-off is a way of computing which aggressively powers down components of computer systems when they are not needed. Simple power gating cannot fully take the chances of power reduction because volatile memories lose data when power is turned off. Recently, new generation non-volatile memories (NVMs) have appeared. Close attention has been paid to "Normally-Off Computing" using these NVMs and cooperation among algorithm, OS, compiler, architecture, circuit and device. In circuit layer, fine grain power managements with non-volatile memory maximize opportunity of power reduction. Since aggressive power management is inseparable from energy and performance overhead, architecture technologies are indispensable to support a wide variety of applications. In software layer, scheduling techniques, which manage activities, are important to maximize energy efficiency.

Based on this strategy, methodology of normally-off computing is introduced. We also include case studies conducted under the Normally-Off Computing project and describe how the real applications are addressed and solved by the methodology. In the project, normally-off computing is applied for three practical applications: healthcare, mobile information devices, and sensor network systems for smart city applications.

Many researchers and students in our project have contributed to the development of normally-off computing. We are sincerely grateful for their great contribution. Our development is also supported by many other researchers through meaningful discussions in our symposia and many other conferences. We also thank anonymous reviewers for their suggestions to make this book better and more purposeful.

Tokyo, Japan

Takashi Nakada
Hiroshi Nakamura

Contents

Chapter 1
Introduction

Takashi Nakada and Hiroshi Nakamura

Abstract Improvement of energy efficiency is indispensable for computer systems in our rapidly evolving information society. To reduce energy consumption without performance degradation, aggressive and careful power management is necessary. Simple power gating cannot fully take the chances of power reduction since volatile memories lose data when power supply is off. However, with new non-volatile memories (NVMs), a synergetic effect for drastic energy reduction is highly expected. Normally-off computing is a way of computing where inactive components of computer systems are aggressively powered off with the help of new NVMs. As a result, high attention has been paid to normally-off computing using these NVMs and cooperation among algorithm, OS, compiler, architecture, circuit and device.

Keywords Power management · Non-volatile-memory · Management granularity

1.1 Background

1.1.1 Sustainable Society

Currently, many kinds of computer systems are working in our life, such as mobile phones, tablet devices and many kinds of sensor devices. Performance requirements of these devices are getting higher year by year. Smart phone is one of representative devices, which provide higher speed mobile connection and sophisticated functions.

T. Nakada (✉) · H. Nakamura
The University of Tokyo, Bunkyo, Tokyo, Japan
e-mail: nakada@hal.ipc.i.u-tokyo.ac.jp

H. Nakamura
e-mail: nakamura@hal.ipc.i.u-tokyo.ac.jp

T. Nakada and H. Nakamura (eds.), *Normally-Off Computing*,
DOI 10.1007/978-4-431-56505-5_1

Additionally, the total number of such devices is also rapidly increased. Several organizations foresee the sensor demand growing to "trillions" by 2017 [2]. This huge number of sensor devices force up the demand of the energy consumption.

To realize sophisticated information society, not only devices but also the entire society requires higher performance and more advanced features. For example, portable devices have a high resolution display. On such devices, contents and user interfaces are also rich and high definition. For data communication, advanced encryption is required to ensure appreciable security. As a result, modern smartphones work only one day on one battery charge and force us to charge them every day.

Therefore, the energy consumption of such information device becomes a serious concern. The total energy consumption limits their operation time. If the operation time of modern devices is less than 24 h, we are forced to recharge them on a daily basis. Their allowable battery capacity is not increased to keep or improve their mobility in terms of size and weight.

From the viewpoint of realizing sustainable society, low energy consumption is strongly desired. One of the ultimate goals is to operate through energy harvesting, which captures electric energy from the environment such as sun light, vibration, radio signal and so on.

From a macro-perspective, the energy consumption of information and communication technology (ICT) accounts for a significant proportion of the total energy consumption in the world. According to a target of the second commitment period in Kyoto Protocol, the total amount of emissions of the six greenhouse gases in Japan should be reduced by at least 25% compared to that of 1990 in 2020.The amount of the carbon dioxide (CO_2) emission in Japan is shown in Fig. 1.1. Meanwhile, in 2012, the amount of emission of the CO_2 is +2.4% compared with that of 1990, the Japanese government estimates that if we don't take any measures to cope with this situation, the amount of emission of the CO_2 will be +4.1%. The government also targets to keep the amount of the CO_2 in 2020 below the level observed in 2012.

Additionally, ICT can help to reduce energy consumption of our society. For example, electrical commerce can help paperless society. Cloud computing can improve energy efficiency by centralized power management. In 2012, the reduction of emission of the CO_2 is -5.4% compared with that of 1990. The government expects this reduction will be -7.5% without any measures. The target of the government is -12.3% by further measures in 2020 [1]. As a result, the ICT will reduce the emission of the CO_2 by about -10% in 2020 and help to reach the target of Kyoto protocol.

Improving energy efficiency is an urgent issue to realize comfortable and sophisticated information society. As power consumption of VLSIs dominates significant portion of the total power consumption of computer systems, low-power techniques for VLSIs are greatly required.

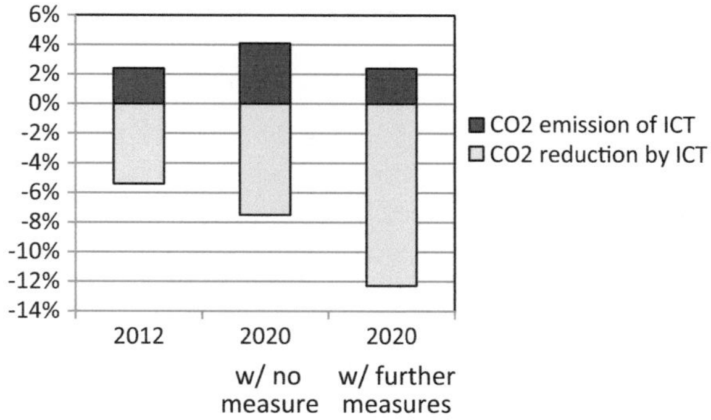

Fig. 1.1 Reduction of CO_2 emission of/by ICT

1.1.2 Power Management

As the power consumption of computer systems is dominated by VLSIs, low power techniques for VLSIs are highly necessary. This power consumption is classified into "dynamic" and "static" power. The former is caused by switching activities of transistors and essentially consumed by computing. On the other hand, the latter is caused by leakage current and always consumed whenever power is supplied.

As VLSI technology improves, the total number of transistors increases. Since each transistor becomes smaller and work with lower voltage, dynamic power consumption of each transistor also becomes smaller. On the other hand, static power consumption of each transistor hardly becomes smaller. As a result, static power increases more rapidly than dynamic power [5], and now it gets comparable to dynamic power consumption. As static power is consumed without any contribution to computing, its reduction is strongly required.

A wide variety of power reduction techniques has been proposed and realized, including Dynamic Voltage and Frequency Scaling (DVFS), clock gating, power gating and so on.

1.1.2.1 Dynamic Power

To reduce the dynamic power, DVFS is one of the most popular techniques. In general, to realize higher performance, larger power consumption is required. Based on this fact, DVFS technique trades performance and power and provide a knob that can adjust the trade-off point. If there exist components which are not bottlenecks in performance, they should be turned into low performance but low power mode by the knob. Then, total power consumption becomes smaller. In addition to this, when the performance is half, the power consumption is less than half. Then, total energy consumption, which is the product of the power and the time, also becomes smaller.

One important observation is that the total energy consumption is the sum of the energy consumption of each component but the performance can limit by a component which is the bottleneck.

Therefore, any components that are not bottlenecks of performance should be switched to lower performance mode. Then lower energy consumption is realized without any performance degradation.

If there is nothing to do, namely during idle period, Clock Gating (CG) can suppress the dynamic power completely in sequential circuits. When CG is applied to a component, clock signal is stopped entirely in that area. The component cannot work anymore, but no dynamic power is consumed.

1.1.2.2 Static Power

Power gating (PG) is a promising way to reduce static power and widely used. When components need not work, PG is applied to shut down their power supply. Then both dynamic and static power consumption becomes almost zero. In modern computer system, all the components need not work all the time during computation. Based on this observation, there exist many chances for PG.

So far, power gating is applied in a coarse manner. Conventional PG requires large overhead in term of both energy and performance. To avoid these large overheads, conventional PG can be applied to a large region and only when very long idle periods are obviously expected.

Recently, however, fine-grain power gating receives much attention because finer granularity increases the chances of PG.

1.1.2.3 Limitation of Power Gating

However, there remain problems to be solved for further power reduction by power gating. Currently, volatile memories are used in traditional VLSIs. When power is turned off, contents or values in the volatile memories are lost. Then, systems cannot resume computation. To avoid this problem, existing power management (1) doesn't power off the volatile memories (Fig. 1.2a) or (2) save/restore the contents to/from an external non-volatile memory for every power gating cycle (Fig. 1.2b). For the method (1), the power reduction is limited due to remaining static power of volatile memory. For the method (2), the power and the performance overheads for save and restore decrease its effectiveness.

To realize optimal power management and make full use of PG, non-volatile memory which consumes lower energy and provides faster access is desired.

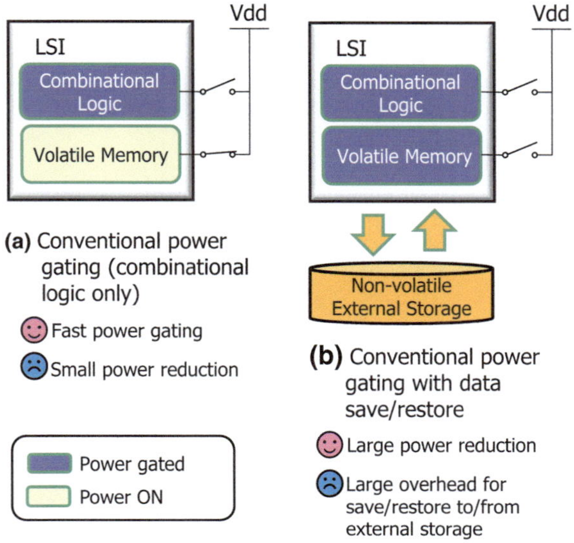

Fig. 1.2 Conventional power management and contents save/restore

1.1.3 Emerging Technology: Non-volatile Memory

Recently, new generation non-volatile memories with new materials become available. These non-volatile memories are more than 1,000× faster than conventional non-volatile memory such as NAND Flash memory. Their access speed is comparable to that of SRAM or DRAM. Additionally, their access energy is much smaller than conventional non-volatile memories/storages such as FLASH or HDD/SSD.

In addition to these advantages, these new generation non-volatile memories have good compatibility with CMOS process. Therefore they are strongly expected to replace the on-chip SRAM and DRAM. The details of these non-volatile memories are introduced in Chap. 3.

These non-volatile memories are good candidate to realize optimal power management. However, these non-volatile memories consumes slightly larger access energy especially write access and slightly longer access latency than volatile memory. To maximize synergetic effect of PG and non-volatile memory, careful consideration is necessary.

1.2 Normally-Off Computing

Normally-off computing is a way of computing where inactive components of computer systems are aggressively powered off by PG with the help of new non-volatile memories.

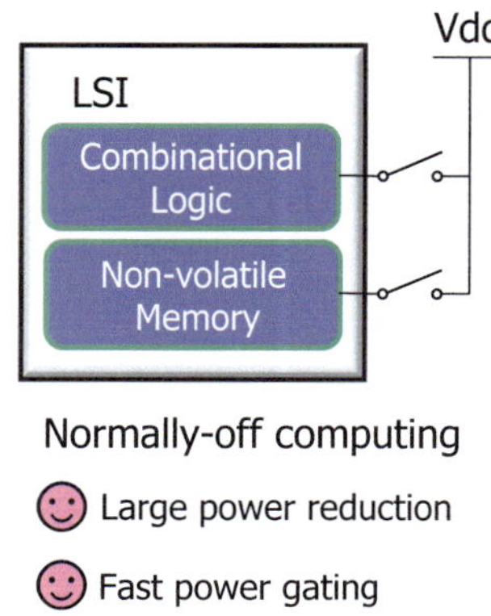

Fig. 1.3 Power management for normally-off computing

For instance, when PG is applied to the volatile memory, it is required to save/restore the contents. In conventional systems, since the external non-volatile memory is slow (Fig. 1.2b), this management is usually handled by system software. The non-volatile memory, however, is quite fast and may be implemented on the same package as cores (Fig. 1.3). Then, management should be done by hardware. In this way, it should be carefully considered when and which contents should be saved/restored and who manages it.

To realize normally-off computing, the most important problems to be considered are (1) Management granularity and (2) organization of memory hierarchy with the non-volatile memory. These two issues are essentially related to their non-volatility.

1.2.1 Management Granularity

1.2.1.1 Spatial Granularity

PG is applied for predesigned area, which is called power domain. To apply PG, all components in a power domain must be idle. Therefore, to maximize chance of PG, power domain should be finer. On the other hand, too small power domain causes huge management overhead due to large number of power domains and each power domain needs their own power switches. As a result, the granularity of power domain should be carefully considered.

1.2.1.2 Temporal Granularity

The principle of the next-generation non-volatile memory is not based on capacitive phenomena, which is completely different from conventional volatile memory. Generally speaking, their leakage power is drastically reduced, but required energy

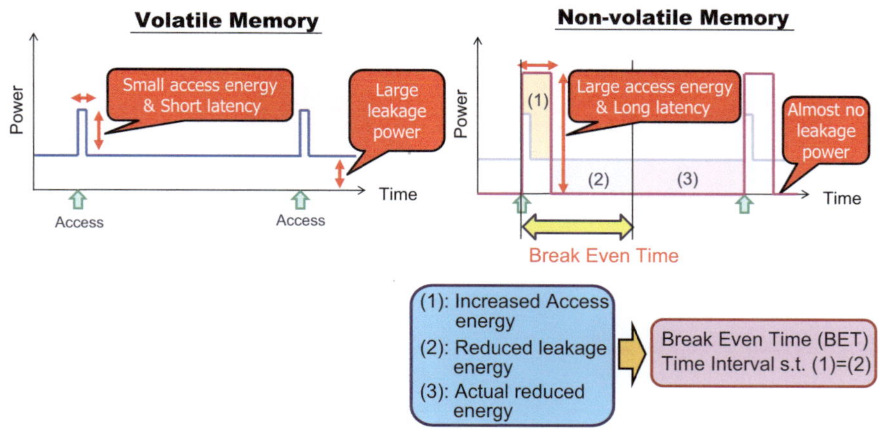

Fig. 1.4 Access energy and Break Even Time (BET)

for access, especially for write access, is higher compared with capacitive volatile memory.

Therefore, to replace the volatile memory with the non-volatile memory, the Break Even Time (BET) must be considered as shown in Fig. 1.4. This figure illustrates power consumption of volatile and non-volatile memories. Volatile memory consumes fairly large leakage power at all the time but its access energy is small. Non-volatile memory, however, consumes large energy when accessed, though its leakage power is almost zero. Therefore, there exists a certain time interval between two consecutive accesses where the increased access energy (1) gets equal to the reduced leakage energy (2). The length of this time interval is defined in Break Even Time or BET. If the interval between two consecutive accesses is longer than the BET, then the power consumption memory is successfully reduced by using non-volatile memory, and the energy saving is represented by (3) in Fig. 1.4. On the other hand, if the interval is shorter than the BET, then use of non-volatile memory leads to the increase of the total power consumption.

In order to reduce the total power consumption, it is indispensable to control the access frequency and keep the access interval longer than the BET. The BETs of non-volatile memories differ depending on their physical mechanisms, and are not the same for read and write accesses. Thus, the access interval, or the temporal granularity of memory accesses, should be carefully controlled by memory access scheduling. This optimization on temporal granularity is a hard problem and cannot be solved without the cooperation among algorithm, OS, compiler, architecture, circuit and device.

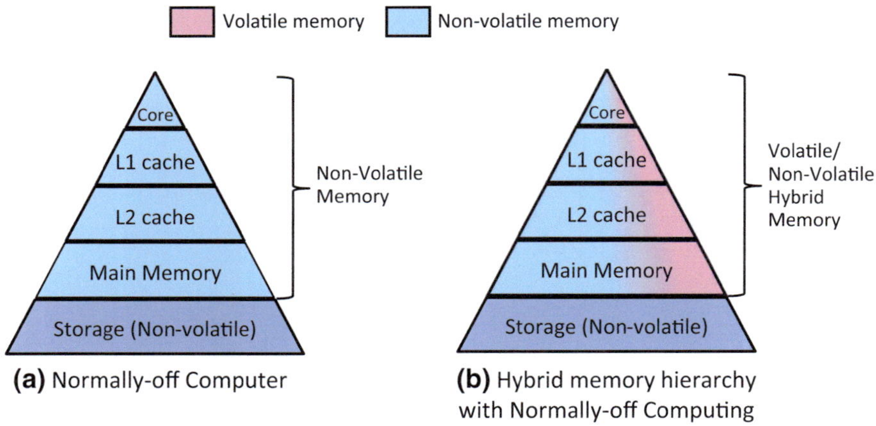

Fig. 1.5 Noff computer and Noff computing

1.3 Expectation on Normally-Off Computing

Originally, "Normally-Off Computer" is featured [3, 4], when new generation non-volatile memories are revealed. If ideal non-volatile memory is available, by combining with fine-grained power gating, normally-off computer consumes energy only when computing is required as shown in Fig. 1.5a.

Unfortunately, current non-volatile memory is still far from the ideal one and the power gating also incurs energy overheads.

"Normally-Off Computing" is a way of computing which aggressively powers off components of computer systems when they need not to operate. The new generation non-volatile memories are very fast and keep its contents without power supply. Thus, by making use of these memories, very fine-grain power control, that is normally-off computing, would become available.

To realize ultimate low-power computing, an extensive viewpoint is important. Therefore, collaboration of hardware, software, middle-ware and architecture is definitely important.

First of all, it is important to understand characteristics of each component and each low-power techniques. There is no ideal solution. Any solution has both advantages and drawbacks. Then, to take their advantage and to eliminate their drawbacks, novel system level design methodology is required. For example, as write energy of non-volatile memory is larger than that of volatile memory, the number of write accesses should be minimized. Such optimization can be considered within multiple layers, such as a write buffer, memory allocation optimization, task scheduling optimization and so on as shown in Fig. 1.5b.

1.4 Organization of This Book

The rest of this book is organized as follows. In Chap. 2, we introduce details of low-power technologies. In Chap. 3, characteristics of several new generation non-volatile memories are summarized. In Chap. 4, normally-off computing architecture is introduced. In Chap. 5, important technologies for realizing normally-off computing are described. In Chap. 6, some practical implementations are explained. In Chap. 7, related research and development are summarized. We conclude in Chap. 8.

References

1. http://www.soumu.go.jp/main_content/000065258.pdf
2. Need for a trillion sensors roadmap. Tsensors Summit. http://www.tsensorssummit.org/Resources/TSensors%20Roadmap%20v1.pdf
3. Ando, K., Yakushiji, K., Kubota, H., Fukushima, A., Yuasa, S., Kai, T., Kishi, T., Shimomura, N., Aikawa, H., Yoshikawa, M., Nagase, T., Nishiyama, K., Kitagawa, E., Daibou, T., Amano, M., Takahashi, S., Nakayama, M., Ikegawa, S., Nagamine, M., Ozeki, J., Watanabe, D., Yoda, H., Nozaki, T., Suzuki, Y., Oogane, M., Mizukami, S., Ando, Y., Miyazaki, T., Nakatani, Y.: Spin-RAM for normally-off computer. In: 11th Annual Non-Volatile Memory Technology Symposium (NVMTS), pp. 1–6 (2011). doi:10.1109/NVMTS.2011.6137104
4. Ando, K., Ikegawa, S., Abe, K., Fujita, S., Yoda, H.: Roles of non-volatile devices in future computer systems: normally-off computers. In: Hu, W.-C., Kaabouch, N. (eds.) Energy-Aware Systems and Networking for Sustainable Initiatives, pp. 83–107. IGI Global, Hershey (2014)
5. Puri, R., Stok, L., Bhattacharya, S.: Keeping hot chips cool. In: 42nd Design Automation Conference, pp. 285–288 (2005). doi:10.1109/DAC.2005.193818

Chapter 2
Low-Power Circuit Technologies

Takashi Nakada

Abstract In this chapter, basic low power techniques are explained including DVFS and other power management such as power gating. Discussion on energy overhead of these techniques is given to understand the limitation on temporal granularity. Normally-off computing tries to apply fine-grain power management with the help of non-volatile memory. Thus, access time and read/write energy of non-volatile memory are also discussed to reveal the condition to reduce energy from the view point of temporal granularity of memory accesses.

Keywords Fine-grain power management · Power gating · DVFS · Break even time

2.1 Introduction

In this chapter, several well-known low-power technologies are explained. Energy consumption is classified into the dynamic and the static energy. The former is caused by switching activities of transistors and essential for computing. On the other hand, the latter is caused by leakage current and not essentially required for computing but consumed whenever power is supplied.

As we briefly introduced in Chap. 1, different types of low power technologies have their own characteristics. For instance, Dynamic Voltage and Frequency Scaling (DVFS) can reduce the dynamic energy by controlling trade-off between performance and energy efficiency. Meanwhile, Power Gating (PG) can eliminate the static power by turning off their power supply while idle state.

In order to realize the optimal energy management, a correlation between low power techniques should be considered carefully. An illustrative example is that of

T. Nakada (✉)
The University of Tokyo, Bunkyo, Tokyo, Japan
e-mail: nakada@hal.ipc.i.u-tokyo.ac.jp

T. Nakada and H. Nakamura (eds.), *Normally-Off Computing*,
DOI 10.1007/978-4-431-56505-5_2

DVFS and PG. In case that DVFS is applied, energy efficiency improves but the execution time gets longer and idle period becomes shorter. Then, the effectiveness of PG is diminished because PG suffers from the transition overhead as described in Sect. 2.3.1.

Modern computer systems consist of many kinds of components and their characteristics is different from one another. One of the most important elements is memory. Flip flops are the most basic memory element and contained in almost all logic circuits. Memory modules are obviously common memory element.

Usually, performance is the most important criteria for many computer systems. To achieve desired performance, high-speed volatile memory is widely used. However, the volatile memories lose their contents when power supply is cut off. Therefore, it is difficult to utilize low-power technologies such as PG.

Recently, new generation non-volatile memories (NV-RAMs) are emerged and comparable performance to the volatile memories. A combination of the NV-RAM and the aggressive power managements present a promising opportunity. A major drawback of the NV-RAM is a write performance. To write information permanently, a longer latency and larger power are required. To achieve significant energy reduction while minimizing performance degradation, how to optimize write operation is critically important.

If volatile RAM is replaced with NV-RAM, stand-by power is not required to maintain its contents but write energy becomes larger. In general, write interval is a key to decide which memory consumes lower energy. Such noticeable parameter that indicates the trade-off time interval is called break even time (BET). However, the amount of write data can be optimized by system configuration such as a memory hierarchy. From a viewpoint of performance, negative impact of the longer write operation can be hidden by a memory hierarchy. Therefore, to find optimal configuration and management, it is important to understand detailed behavior and characteristics of each component and each low power technology.

The first step to realize the optimal power management is to understand the behavior of the target system. To recognize idle periods are essentially important, as the low-power technologies reduce the energy consumption during idle periods.

Idle periods have wide variety of time interval when systems are under operation. Especially, their temporal and spatial granularities are the most important characteristics. The spatial granularity corresponds to management unit in terms of area, such as power domain for PG. If the spatial granularity is finer, the number of components in one area is fewer, then opportunity to switch to low-power mode increases. If the granularity is coarser, the opportunity decreases but control overhead becomes smaller, because the area of control logic is relatively smaller.

In general, hardware implementation can minimize management overhead, but its behavior is given by predefined logic. Conversely, software implementation can manage more flexible, but control overhead is relatively larger. To realize optimal management, collaboration with each other is significantly important.

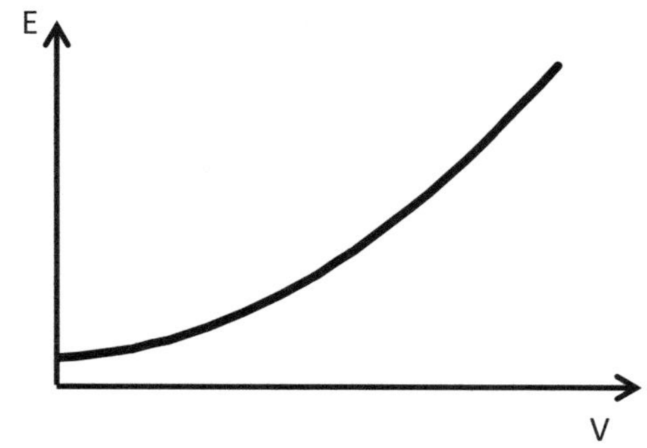

Fig. 2.1 Energy model

2.2 Basics of Low Power Techniques

In this section, we briefly introduce major power reduction techniques including clock gating, power gating, DVFS, and so on.

In general, the relation between energy, voltage and clock frequency can be modeled by following well known equation [5] and shown in Fig. 2.1.

$$E_{proc} = \alpha_1 T_1 C V^2 f + T_2 V I_{leak} \tag{2.1}$$

Here, E_{proc} represents the energy consumption of the microprocessor. α_1, T_1, C, V and f represent a switching activity, the execution time, the circuit capacity, the supply voltage, the operation frequency respectively. T_2 and I_{leak} represent the total time that includes idle period and the leakage current respectively.

The first and the second clauses represent the dynamic and the static energy respectively. The former is caused by switching activities of transistors and essentially consumed by computing. On the other hand, the latter is caused by leakage current and always consumed whenever power is supplied.

DVFS is a popular way to reduce the dynamic power and has been around for more than a decade [10]. DVFS allows the voltage and the clock frequency to be decreased dynamically to trade time for energy.

Clock Gating (CG) simply cuts clock delivery from a clock oscillator. In general, the oscillator keeps running for a quick restart. Since this technique has no performance penalty, when there is no ready task, processor cores should switch to clock gating mode unless other low power techniques are applicable.

Power Gating (PG) is a promising way to reduce the static power. In modern computer system, all the components need not work all the time during computation.

Dynamic Power Management (DPM) [4] manages power states of the components based on the BETs. The power states are defined in each component as a power knob. When an idle period is encountered, the BET of each component is compared with the length of the idle period. If the BET is longer than the length of the idle period, the component should be clock gated or power gated. Therefore, when the length

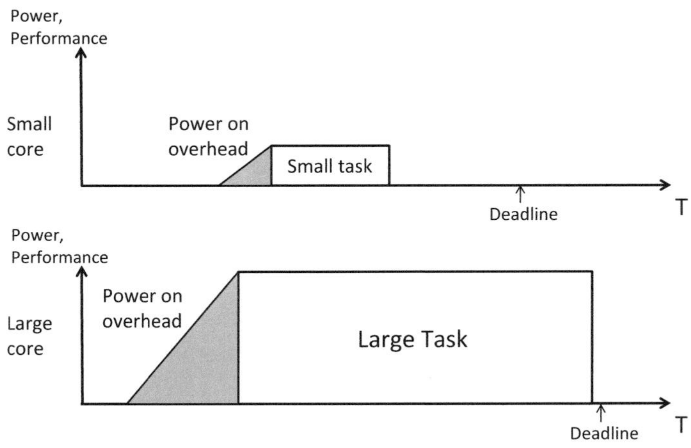

Fig. 2.2 Heterogeneous cores

of idle period becomes longer, more components can be clock gated or power gated and energy reduction is more significant.

For the combination of DPM and DVFS, tradeoffs between the two techniques should be considered [6]. When DVFS is used, the clock frequency is decreased to reduce the energy consumption during the execution of tasks, while the execution time increases and the idle time decreases.

2.2.1 Heterogeneous Hardware

A heterogeneous hardware is a solution to adopt a variety of applications.

Modern mobile devices such as smartphones are integrated with heterogeneous processor such as big.LITTLE architecture [2]. As energy efficiency largely depends on hardware architecture, optimal core selection helps minimizing total energy consumption.

For example, when low performance is required, small core, which is energy efficient, is preferable. Large core is used only when high performance is necessary to meet their requirement. Based on this idea, usage of energy efficient core is maximized and total energy consumption is reduced. Additionally, larger core requires larger power on overhead.

Figure 2.2 shows a simple example of a heterogeneous execution. Y-axis shows relative performance and power consumption. For a small task, small core is suitable to minimize power consumption. For a large task, large core is necessary to meet its deadline even its energy efficiency is low. As a result, it is important to execute as much task as possible should be executed on small core.

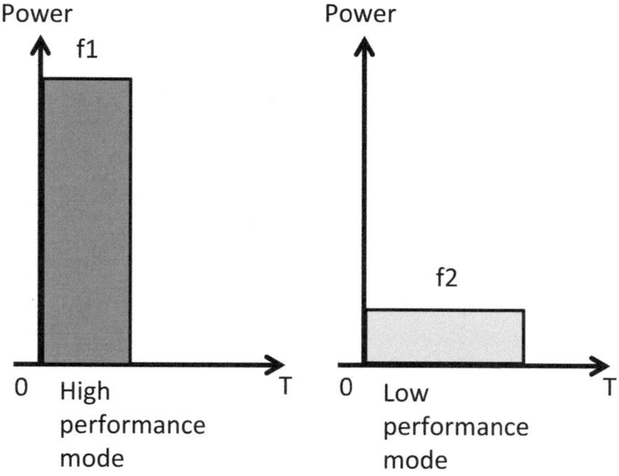

Fig. 2.3 Dynamic Voltage and Frequency Scaling (DVFS)

2.2.2 *DVFS*

As shown in Eq. 2.1, the higher performance is realized by higher voltage, frequency and the larger circuit that causes the larger circuit capacity.

Dynamic Voltage and Frequency Scaling (DVFS) allows the voltage and the clock frequency to be decreased dynamically to trade time for energy. A lot of research is done in this area. By considering the consumed energy as a cost function, while considering deadlines as constraints, a mathematical problem can be defined and the optimal clock frequencies can be found for many kinds of real-time systems [7, 11]. For the combination of DPM and DVFS, tradeoffs between the two techniques should be considered [6]. When DVFS is used, the clock frequency is decreased to reduce the energy consumption during the execution of tasks, while the execution time increases and the idle time decreases.

Figure 2.3 shows executions of same task with two different performance modes. In high performance mode (left), the execution time is shorter but the power consumption is higher. On the other hand, low performance mode needs longer execution time but lower power consumption. According to Eq. 2.1, the total energy consumption, which is represented by the area in the figure, of high performance mode is larger than that of low performance mode. Thus, DVFS can trade between the performance and the energy consumption.

Additionally, DVFS and heterogeneous processors are complementary to each other. DVFS can adjust performance finely but adjustment range is limited. On the other hand, heterogeneous processors can adjust performance largely but possible performance mode is limited by the number of implemented cores. Their combination realizes fine and wide performance adjustment.

2.2.3 Dynamic Power Management

An overview of Dynamic Power Management (DPM) techniques is given in a survey article [4]. DPM manages both static and dynamic power dynamically. Thus DPM is important to reduce the static power when the processor core is in an idle state. This is in contrast with DVFS, which mainly reduces the dynamic power.

An example of a typical set of power modes is shown in Fig. 2.4 and Table 2.1. In active mode, all components are turned on. For a short idle interval, Clock Gating (CG) is preferable. The processor core can restart from the CG state instantly. In Core Power Gating mode, PG is applied to the processor core to obtain more energy reduction. To return from this mode, several clock cycles are required and some transition energy overhead is consumed. Vcc power gating is applied for a very long idle period. In this mode, the power supply is completely cut off. The only way to recover from this mode is to resume power supply. Then the processor core should follow almost the same as a power on reset procedure. Therefore, we can get the largest power reduction but both time and energy overhead become the most costly.

As mentioned in the previous section, there exist BETs between these power modes. To determine the appropriate power state, the length of next idle period is compared with these BETs. This strategy can be modeled with a function of cost with the length of the idle period. This function turns out to be piecewise-linear, increasing and concave [3].

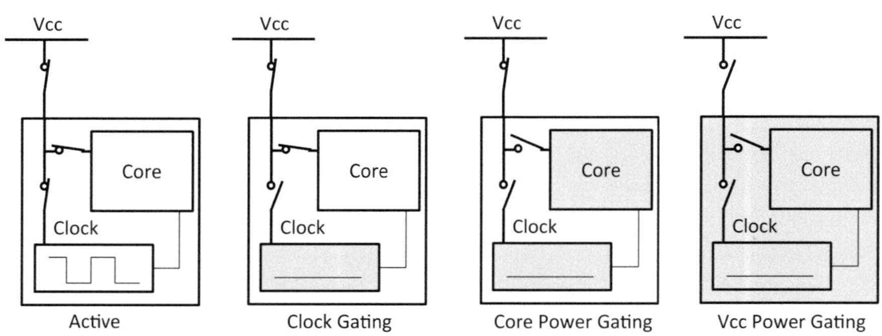

Fig. 2.4 Power modes

Table 2.1 An example of power mode

	Vcc	Core	Clock
Active	ON	ON	ON
Clock gating	ON	ON	OFF
Core power gating	ON	OFF	OFF
Vcc power gating	OFF	OFF	OFF

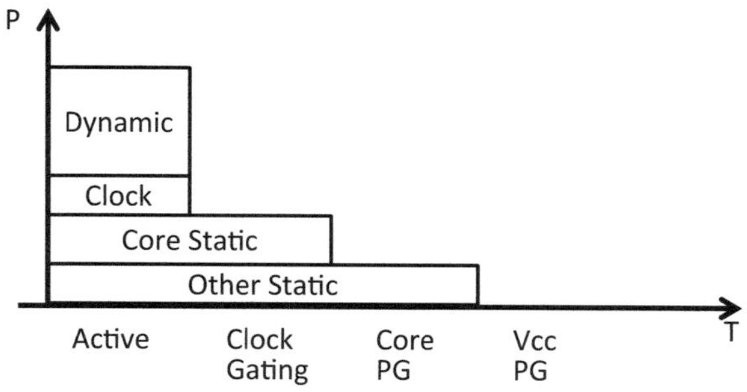

Fig. 2.5 Power breakdown

Figure 2.5 shows power breakdown of power modes. In active state, all components are on. When clock gating is applied, the processor core is stopped and no execution is possible but dynamic power of processor core and clock logic is suppressed. When core PG is applied, static power is reduced additionally. Finally, when Vcc PC is applied any power is not consumed except stand-by power of power supply which is out of scope of this book.

In embedded systems, executed tasks are fixed and periodic and their scheduling is known. So, when an idle state is encountered, the time when the next task can be invoked is definitely predictable. Then the length of the idle period is also predictable. Additionally, since the restart time is predictable, wakeup overheads are easily hidden by a pre-wakeup technique. Therefore, the optimal power management is easily determined by the strategy. Finally, DPM related parameters include hardware parameters and the length of the idle period.

2.2.4 Sleep Mode

As an extension of DPM, modern hardware has more low power modes, called sleep modes or sleep states. Sleep modes are defined for each component. For example, the Advanced Configuration and Power Interface (ACPI) specification provides an open standard for device power management. First released in December 1996, ACPI defines platform-independent interfaces for power management, monitoring and related technologies.

2.2.4.1 Global/Processor State

The ACPI [1] specification defines four Global "Gx" states as shown in Table 2.2. G1 (Sleeping) is divided int four Sleep "S1–S4" states and G2 (Soft Off) also defined as "S5" as shown in Table 2.3.

In both state definitions, the larger number corresponds to the deeper sleep mode.

Table 2.2 Global system state definition (cited from [1])

State	Definition
G0 working	A computer state where the system dispatches user mode (application) threads and they execute
G1 sleeping	A computer state where the computer consumes a small amount of power, user mode threads are not being executed, and the system "appears" to be off. Latency for returning to the Working state varies on the wake environment selected prior to entry of this state
G2 soft off	A computer state where the computer consumes a minimal amount of power. No user mode or system mode code is run. This state requires a large latency in order to return to the Working state
G3 mechanical off	A computer state that is entered and left by a mechanical means. It is implied by the entry of this off state through a mechanical means that no electrical current is running through the circuitry

The CPU power states "Cx" are defined as shown in Table 2.4. Additional states are defined by manufacturers for some processors. For example, Intel's Haswell platform has states up to C10, which defines core states and package states.

Basically, deeper sleep state consumes less static power, but consumes more transition energy and takes longer transition time. In general, the length of each idle period is hard to predict. To realize an adaptive control, these states are managed by a timeout-based scheme. When an idle time continues more than predefine threshold times, their state will be deeper sleep mode.

If the length of an idle period is predictable, optimal sleep state can be determined and transit to the state directory to minimize overheads. For longer idle periods, deeper sleep states are preferable. Therefor the length of idle period is important to realize optimal power management. Additionally, there exists a specific length of idle period between two adjacent sleep modes. That length is definitely the BET for these sleep modes.

2.2.5 *Fine-Grained Power Gating*

Power gating (PG) is a promising way to reduce static power and used mainly in embedded systems. In modern computer systems, all the components need not work all the time during computation. Based on this observation, there exist many chances for PG. So far, PG is applied in a coarse manner. Especially its temporal granularity

Table 2.3 Sleeping state definition (cited from [1])

State	Definition
S1 sleeping state	The S1 is a low wake latency sleeping state. In this state, no system context is lost (CPU or chip set) and hardware maintains all system context
S2 sleeping state	The S2 is a low wake latency sleeping state. This state is similar to the S1 sleeping state except that the CPU and system cache context is lost
S3 sleeping state	The S3 is a low wake latency sleeping state where all system context is lost except system memory. CPU, cache, and chip set context are lost in this state. Hardware maintains memory context and restores some CPU and L2 configuration context
S4 sleeping state	The S4 is the lowest power, longest wake latency sleeping state supported by ACPI. In order to reduce power to a minimum, it is assumed that the hardware platform has powered off all devices. Platform context is maintained
S5 soft off state (G2 soft off)	The S5 state is similar to the S4 state except that the OS does not save any context. The system is in the "soft" off state and requires a complete boot when it wakes. Software uses a different state value to distinguish between the S5 state and the S4 state to allow for initial boot operations within the BIOS to distinguish whether or not the boot is going to wake from a saved memory image

is quite coarse. Recently, however, fine-grain PG receives much attention because finer granularity increases the chances of PG.

PG is a representative static power reduction technique, which helps cutting off the power supply to idle circuit blocks by turning off (or on) the power switches which are inserted between the GND/VDD lines and the blocks. PG has been applied to different types of circuit blocks with various granularities.

So far, power gating is applied in a coarse-grain manner. Recently, however, fine-grain power gating receives much attention because finer granularity increases the chances of PG. For example, Geyser-3 [8, 9] implements a fine-grained run-time PG. In these processors, PG is applied to function units (FUs). Each FU can be powered on or off instruction by instruction. In other words, instruction-level power gating is implemented in these processors. Based on this observation, there exist many chances for PG in a wide range of idle periods.

Table 2.4 Processor power state definition (cited from [1])

State	Definition
C0 processor power state	While the processor is in this state, it executes instructions
C1 processor power state	This processor power state has the lowest latency. The hardware latency in this state must be low enough that the operating software does not consider the latency aspect of the state when deciding whether to use it. Aside from putting the processor in a nonexecuting power state, this state has no other software-visible effects
C2 processor power state	The C2 state offers improved power savings over the C1 state. The worst-case hardware latency for this state is provided via the ACPI system firmware and the operating software can use this information to determine when the C1 state should be used instead of the C2 state. Aside from putting the processor in a non-executing power state, this state has no other software-visible effects
C3 processor power state	The C3 state offers improved power savings over the C1 and C2 states. The worstcase hardware latency for this state is provided via the ACPI system firmware and the operating software can use this information to determine when the C2 state should be used instead of the C3 state. While in the C3 state, the processor's caches maintain state but ignore any snoops. The operating software is responsible for ensuring that the caches maintain coherency

Generally speaking, coarser-grain PG can reduce more static power but have a larger transition overhead. Thus, coarse-grain PG should be applied only for long idle period. For a short idle period, finer-grain PG is preferable. Therefore, there exists certain idle time between PGs. The boundary times are called BETs.

Granularity is a key metric to understand what the best power management is. In power gating technology granularity corresponds to a power domain. In a same power domain, power supply is controlled together by one control signal. Namely, a finer power domain controls power supply of a smaller area. The power domains can be hierarchical, namely a coarser power domain can contain several finer power domains. When the coarser power domain turns off, all of the finer power domains in it necessarily turn off.

A simple example is shown in Fig. 2.6. In this example, gray color shows power off area and others are on. In left processor, only one function unit is turned off by fine-grain PG but right processor is totally powered off by coarse-grain PG.

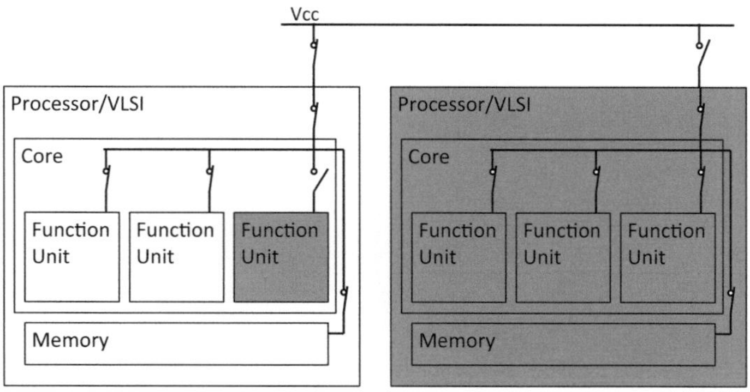

Fig. 2.6 Hierarchy of power domain

From view point of circuit, in general, to minimize control overhead it is better to use as coarse domain as possible. Because, to realize fine grain power management, a large number of power switches and control logics are required. The static power of the switched and logics cannot be turned off by themselves. On the other hand, when the power domain is larger, the opportunity of PG may be shorter in time domain. Therefore, to realize optimal power management, hierarchical power domain management, which is combined coarse and fine domain managements.

With such hierarchical management, as large as possible power domain should be turned off. Then within the remaining domain, finer power management should be applied. When a coarser power domain turns off, several finer domains in it are automatically turns off with their power switches and control logics.

As a result, it is important that as coarse as possible power domain should be powered off for as long time period as possible.

2.3 Energy/Performance Trade-Off

In this section, we take a general view of energy overheads of low power technologies. There are two type of overhead; one is an energy overhead, the other is a performance overhead. As the performance overhead can be hidden and largely depends on application and other situations, we focus on the energy overhead.

We formulate typical trade-offs and clarify important parameters to achieve energy reduction.

As we have already introduced several low-power technologies, there may exist several low-power modes for a particular control domain. There are trade-offs between these modes. Namely, lower power mode can reduce more static power but need larger transition overhead. If this is not true, it means that there exist a low-power mode that is more energy efficient and needs lower overhead than other

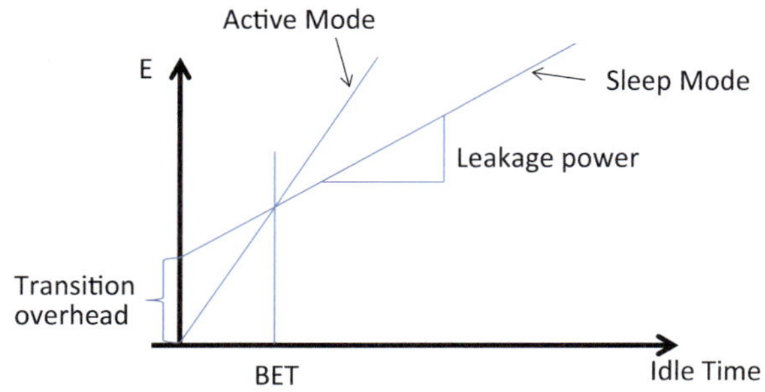

Fig. 2.7 Energy model and BET

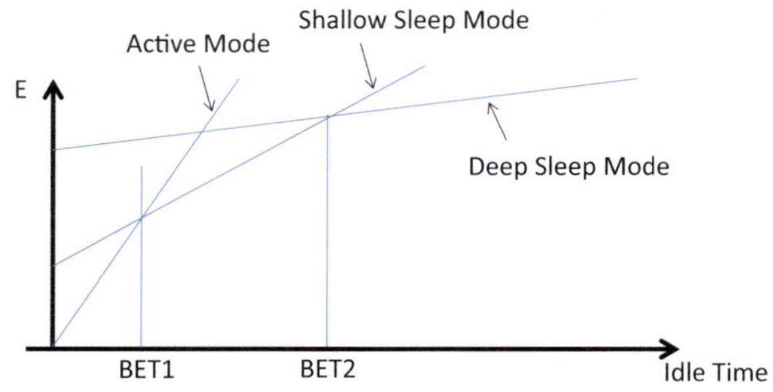

Fig. 2.8 BETs with multiple sleep modes

modes. Then, we can always use such ideal low-power mode. In the real systems, such ideal mode is not exits. Therefore, an adaptive power management is important.

2.3.1 *Transition Energy and BETs*

When a component has only one low power mode, to achieve energy reduction by this power mode, the minimum length of an idle period *BET* is given by follows.

$$BET = E_{OH}/(P_{active} - P_{sleep}) \tag{2.2}$$

Here, E_{OH} represents transition overhead energy. P_{active} and P_{sleep} represent power consumptions when active and low power states respectively. If the length of idle period is longer than the BET, that component should turn into low power mode to reduce total energy consumption.

Figure 2.7 shows energy models of active mode and sleep mode. As there is no transition overhead for active mode, the energy function of active mode starts from origin. On the other hand, as there is some transition overhead for sleep mode, the energy function of sleep mode starts from higher than origin. Meanwhile, the energy function of sleep mode has more moderate sloop than that of active mode. Therefore, there are cross point and we call it BET. If the idle time is longer than the BET, sleep mode is preferable. Otherwise, sleep mode should not be chosen.

When a component has multiple low power modes, *BET* can be extended as follows.

$$BET_i = (E_{OH_i} - E_{OH_{(i-1)}})/(P_{(i-1)} - P_i) \quad (2.3)$$

Here, the component has N modes and transition overhead energy and power consumption of ith mode are given by E_{OH_i} and P_i respectively. The energy functions are shown in Fig. 2.8. We assume $E_{OH_i} > E_{OH_{(i-1)}}$ and $P_i < P_{(i-1)}$, namely, 1st mode is the deepest sleep mode and Nth mode is the shallowest. As 0th mode corresponds to an active state, $E_{OH_0} = 0$ and $P_0 = P_{active}$.

When BET_i is longer than the length of idle mode, ith sleep mode can reduce total energy consumption. If multiple BET_i satisfy this condition, deeper sleep mode can reduce more energy.

In this example, if the length of idle period is longer than BET_2, deep sleep mode should be chosen. If the length of idle period is shorter than BET_1, it is better to stay active mode. In any other case, shallow sleep mode should be chosen.

Note, this discussion assume the component directly transit to the suitable sleep mode from its active state. When the transitions occur step by step, BET_i will be longer.

2.3.2 Access Energy of NVRAM and BET

For the access energy of NVRAM, same discussion of low power mode is possible. The latest NVRAM consumes less static energy but consumes larger access energy especially for write operation as shown in Fig. 2.9. To simplify this discussion, we assume the read access energy of NVRAM is same as that of volatile memory and the static energy of NVRAM is 0. Then, to realize energy reduction, the average interval of the write accesses BET_{NV} is given by follows.

$$BET_{NV} = (WE_{NV} - WE_V)/P_V \quad (2.4)$$

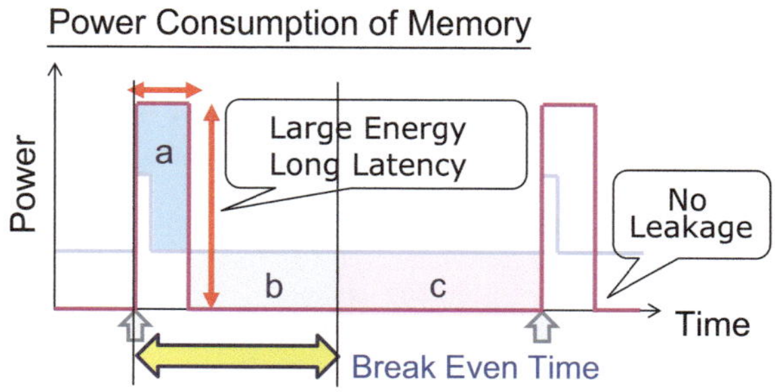

Fig. 2.9 BET in NVRAM

Here, WE_{NV} and WE_V represent write energy per write access for NVRAM and volatile memory respectively. P_V represents the static power of the volatile memory. When the average write interval is longer than BET_{NV}, NVRAM can reduce total energy consumption.

2.4 Summary

Low power technologies of computer systems are indispensable for the forthcoming sustainable and sophisticated information society. Normally-off computing is one of the promising ways to achieve this goal. In this paper, we described its expectation and introduced our Normally-Off Computing project.

In this chapter, we attempt to find the best way to make full use of the new generation non-volatile memories. So far, from the view point of memory system optimization, only two properties of memory are focused on and discussed, that is latency and capacity. Now, non-volatility appears as the third property.

This property absolutely contributes further power reduction. However, it will not effectively lead to power reduction without careful considerations because of two major problems or challenges, that is, memory hierarchy and temporal granularity. In other words, once we overcome these problems, further power reduction would be achieved by using dreamy zero-leakage non-volatile memories. Cooperation and cooptimization of different design layers, including algorithm, OS, compiler, architecture, circuit and device, are definitely required.

Currently a lot of researches on new memory devices are performed. Other types of useful memory devices other than what introduced in this paper may become available in the future. However, it would not be fruitful if we discuss the device properties only. The viewpoint of "computing technology," the technology how to make full use of the attractive property is essentially important.

References

1. Advanced configuration and power interface specification 5.0a. http://www.acpi.info/DOWNLOADS/ACPI_5_Errata%20A.pdf
2. Arm unveils its most energy efficient application processor ever; redefines traditional power and performance relationship with big.little processing (Press release). ARM Holdings (2011)
3. Augustine, J., Irani, S., Swamy, C.: Optimal power-down strategies. In: 45th Annual IEEE Symposium on Foundations of Computer Science, pp. 530–539 (2004). doi:10.1109/FOCS.2004.50
4. Benini, L., Bogliolo, A., De Micheli, G.: A survey of design techniques for system-level dynamic power management. IEEE Trans. Very Large Scale Integr. (VLSI) Syst. **8**(3), 299–316 (2000). doi:10.1109/92.845896
5. Burd, T., Brodersen, R.: Energy efficient cmos microprocessor design. In: Twenty-Eighth Hawaii International Conference on System Sciences, vol. 1, pp. 288–297 (1995). doi:10.1109/HICSS.1995.375385

6. Gerards, M.E.T., Kuper, J.: Optimal dpm and dvfs for frame-based real-time systems. ACM Trans. Archit. Code Optim. **9**(4), 41:1–41:23 (2013). doi:10.1145/2400682.2400700. http://doi.acm.org/10.1145/2400682.2400700
7. Huang, W., Wang, Y.: An optimal speed control scheme supported by media servers for low-power multimedia applications. Multimed. Syst. **15**(2), 113–124 (2009). doi:10.1007/s00530-009-0153-5
8. Kondo, M., Kobyashi, H., Sakamoto, R., Wada, M., Tsukamoto, J., Namiki, M., Wang, W., Amano, H., Matsunaga, K., Kudo, M., Usami, K., Komoda, T., Nakamura, H.: Design and evaluation of fine-grained power-gating for embedded microprocessors. In: Proceedings of the Conference on Design, Automation & Test in Europe, DATE 2014, pp. 145:1–145:6. European Design and Automation Association, 3001 Leuven, Belgium (2014)
9. Usami, K., Kudo, M., Matsunaga, K., Kosaka, T., Tsurui, Y., Wang, W., Amano, H., Kobayashi, H., Sakamoto, R., Namiki, M., Kondo, M., Nakamura, H.: Design and control methodology for fine grain power gating based on energy characterization and code profiling of microprocessors. In: 19th Asia and South Pacific Design Automation Conference (ASP-DAC), pp. 843–848 (2014). doi:10.1109/ASPDAC.2014.6742995
10. Weiser, M., Welch, B., Demers, A., Shenker, S.: Scheduling for reduced cpu energy. In: Proceedings of the 1st USENIX Conference on Operating Systems Design and Implementation, OSDI 1994. USENIX Association, Berkeley, CA, USA (1994). http://dl.acm.org/citation.cfm?id=1267638.1267640
11. Yao, F., Demers, A., Shenker, S.: A scheduling model for reduced cpu energy. In: 36th Annual Symposium on Foundations of Computer Science, pp. 374–382 (1995). doi:10.1109/SFCS.1995.492493

Chapter 3
Non-volatile Memories

Koji Ando, Shinobu Fujita, Masanori Hayashikoshi and Yoshikazu Fujimori

Abstract This chapter describes the basic properties of various computer memories. Historical evolution of the roles of non-volatile functionalities in computer architecture is discussed to explain recent rejuvenating interest in new non-volatile memory technologies. Next, required properties for memories, such as scalability, access speed, power consumption, are discussed referring to those of current main-stream memories, i.e., dynamic random access memory (DRAM), static RAM, and NAND flash memory. Then, histories, working principles, and properties of spin-transfer torque magnetoresistive RAM, resistive RAM, phase change RAM, ferroelectric RAM, and NOR flash memory are described. Finally, possible positioning of these various non-volatile memories in future computer architecture are discussed.

Keywords Non-volatile memories · MRAM · ReRAM · PCRAM · FeRAM · NOR flash memory

K. Ando (✉)
National Institute of Advanced Industrial Science and Technology, Tsukuba, Ibaraki, Japan
e-mail: ando-koji@aist.go.jp

S. Fujita
Toshiba Corporation, Minato, Tokyo, Japan
e-mail: shinobu.fujita@toshiba.co.jp

M. Hayashikoshi
Renesas Electronics Corporation, Koto, Tokyo, Japan
e-mail: masanori.hayashikoshi.cj@renesas.com

Y. Fujimori
ROHM Co., Ltd., Kyoto, Japan
e-mail: yoshikazu.fujimori@mnf.rohm.co.jp

T. Nakada and H. Nakamura (eds.), *Normally-Off Computing*,
DOI 10.1007/978-4-431-56505-5_3

3.1 Introduction

Processing, memory, and storage of information are three fundamental elements of computing. Computers of the 60s (Fig. 3.1a) were composed of a processor with a few-MHz clock cycle, non-volatile magnetic core memory, and magnetic tape storage. The processor was composed of discreet transistors, and its operation speed was slow enough to be well balanced with the performances and capacities of the memory and storage. The computer architecture was simple, and the magnetic core memory was a universal memory with sufficient speed and capacity and non-volatility.

Development of Si integrated circuit technology has disrupted the above balance. Integration of a huge amount of transistors into a single Si chip enabled the rapid enhancement of processor performance and a large reduction in production cost of processor. To rebalance processor-memory performance with this processor performance enhancement and cost reduction, more memory capacity was required. However, the cost of magnetic core memory could not be sufficiently reduced because magnetic materials were not suitable for integration at that time. Dynamic random access memory (DRAM), which was invented in the late 60s, is composed of one transistor and one capacitor (Fig. 3.2), and can be integrated using Si integrated

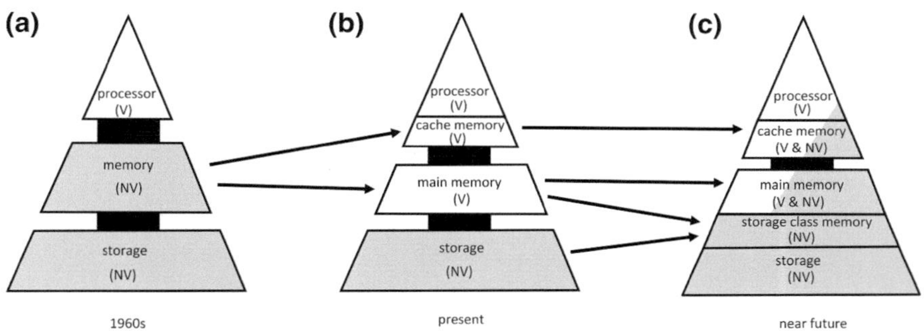

Fig. 3.1 **a** Once, computers were composed of non-volatile storage, non-volatile memory, and volatile processors. **b** Now, memory is split into main memory and cache memory, both of which are volatile. **c** In near future, non-volatility is expected to be re-introduced into memory layers and even into processors

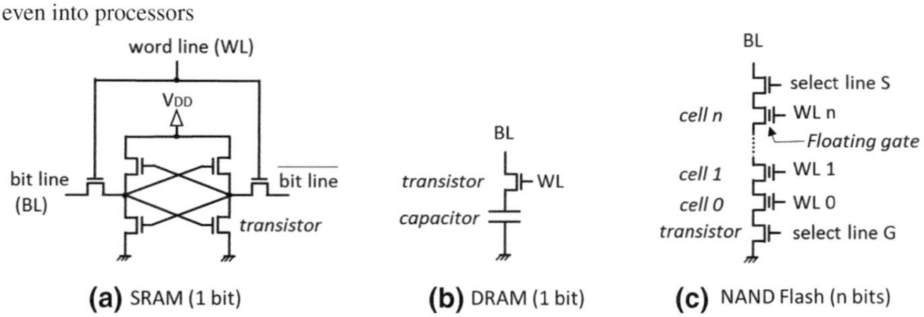

Fig. 3.2 Structures of SRAM, DRAM, and NAND flash memory

circuit technology. The DRAM has a drawback in that cyclic refresh charging is indispensable to compensate for charge leakage from the capacitor. Furthermore, it is a volatile memory. Nevertheless, the low production cost of DRAM was decisive to drive magnetic core memory out of computer architecture.

However, DRAM performance was not sufficient for processors. The speed of DRAM is limited by the charging and dis-charging processes of an electric capacitor, which is used to store information. Static random access memory (SRAM) (Fig. 3.2), which is a pure transistor circuit to emulate memory function, is used to enable faster memory functionality to bridge the performance-gap between processor and DRAM. This memory architecture composed of DRAM main memory and SRAM cache memory is now used in every computer (Fig. 3.1b). The non-volatility was lost from the memory layer and remains only in the storage layer.

Recently, the introduction of non-volatility into memory layers, even into the processor layer (Fig. 3.1c), has been attracting much attention. This rejuvenated interest in non-volatility is due to (i) fear that the scaling of Si complementary metal-oxide-semiconductor (CMOS) large-scale integration (LSI) technology is now ending, (ii) advancements in non-volatile memory technologies, and (iii) emergence of new type of applications.

For a long time, materials used for Si LSI technology have been limited, e.g., Si, SiO_2, As, B, P, Al, and Cu. Without introducing drastic changes in materials and transistor structure, only reduction in transistor size along a simple guideline called the scaling law [1] has been enhancing transistor performance and increasing the density of integrated transistors. Performances of processor, SRAM, DRAM, and NAND flash memory have been dramatically increased in this way.

In the LSI design, power dissipation of each device has been always a key factor. Because a practically acceptable level of heat generation per unit Si chip area is limited, the power dissipation of each device integrated into LSI should be smaller for larger scale integration with smaller device size. A reason CMOS transistors had expelled bipolar transistors from LSI is COMS's extraordinary low-power dissipation.

The scaling law had been working well for transistors larger than 90 nm. However, for smaller transistors, the validity of the simple scaling law had begun to breakdown; thus, power dissipation cannot be sufficiently reduced anymore. In smaller transistors, isolation between the source and drain degrades, and the leak current flows between the source and drain even under a standby state (sub-threshold leak current). Recently, this meaningless static power dissipation of SRAM has become overwhelming the meaningful active power used for information manipulation. In a typical mobile usage case, it can be as large as 80% the average processor power [2]. Because loading of more cache memories is indispensable to enhance computer performance, new smaller cache memory with smaller static power than SRAM is expected.

The DRAM also suffers from scaling because the required electrical capacitance for its memory function, as large as several tens of fF, has become more difficult for smaller cell size.

The size of NAND flash memory is already as small as 20 nm or less, and the electrons used to store the information can be as few as several hundred. Further scaling is difficult.

All these difficulties that SRAM, DRAM and NAND flash memory are encountering raise strong expectations for new non-volatile memories.

In response to these expectations, a variety of non-volatile memory technologies have been advanced recently. Silicon LSI technologies had been reluctant to accept exotic materials required for non-volatile memories. However, to circumvent the scaling problem, state-of-the-art Si LSI technology has already incorporated a variety of new materials such as Ta_2O_5 and HfO_2. This lowered the barrier for non-volatile memory to be included into the LSI process.

Newly emerging uses of IT instruments also require new non-volatile memory. Mobile instruments that work with batteries require ultra-low power-consuming devices for longer operation. At the same time, many advanced functional circuits should be packed into their tiny bodies. Highly integrated and low-power consuming processors and memories are required. Data servers also require a new class of non-volatile memory called storage-class-memory (SCM). At present, there is an access speed-gap as large as 3 orders of magnitude between DRAM and storage devices, i.e., hard-disk drives (HDDs) and solid-state drives (SSDs). Recent demands for handling huge data require SCM to bridge the access speed-gap between storage and main memory.

In this chapter, we discuss emerging resistor-based and charge-based non-volatile memories. Resistor-based memories are spin-transfer torque magnetoresistive RAM (STT-MRAM), resistive RAM (ReRAM), and phase-change RAM (PCRAM). Charge-based memories are ferroelectric RAM (FeRAM) and NOR flash memory.

The required characteristics for memories are discussed below by referring to typical values of commercially available SRAM, DRAM and NAND flash memory (Fig. 3.2 and Table 3.1).

Scalability

Cell area size per bit is one of the most important parameters of memory because it impacts both cost and performance. A simpler cell design with smaller components is preferred.

Because SRAM requires six transistors (Fig. 3.2a), its cell area is as large as 100–300 F^2 where F is the smallest dimension, i.e., feature size, of CMOS technology. This makes SRAM expensive. The SRAM-based cache capacity is typically limited to 1 kB–32 MB depending on the application.

A DRAM cell (Fig. 3.2b) is composed of one transistor and one capacitor. The cell size of this 1 transistor -1 capacitor (1T-1C) DRAM can be as small as 6 F^2. A 8-Gbit DRAM with $F = 20\,\text{nm}$ is under mass production. The transistor selects the assigned bit from the memory cell array by activating the bit and word lines. The combination of one selecting transistor and one memory element is a commonly adopted in memory cell design because any bit can be randomly accessed. Resistor-based memories also adopt the 1T-1R design. The smaller cell requires a smaller selecting transistor. However, as the size shrinks, the transistor's ability to drive the memory element becomes weak. If the memory element requires large current to write information, a large transistor must be used, and the cell size is determined by the size of the selecting transistor not by the size of the memory element.

Table 3.1 Typical performances of commercially available SRAM, DRAM, and NAND Flash memory

	SRAM	DRAM	NAND
Scalability	100–300 F^2, 8 MB F = 14 nm	6 F^2, 8 GB@F = 20 nm	2–4 F^2, 128 GB@F = 15 nm
Typical memory capacity (2015)	8 kb–256 Mb	4–8 Gb	128–256 Gb
Random access time	10 ps–5 ns	10–40 ns	10 us (read) 500 us (write) 1 ms (erase)
Access power	1–10 uW/bit	100–200 nW/bit	Several W/page
Static power	10 p–100 nW/bit	1 nW/bit	0
Random accessibility	○	○	×
Non-volatility	×	×	○
Retention	∞@powered	∞@powered	10 years<
Endurance	∞	∞	10^2–10^6
Temperature	–40 to 125 °C	–40 to 125 °C	–40 to 85 °C

(Commercial products)

The cross-point memory array architecture (Fig. 3.3) is effective in reducing the cell size. At the cross point between the orthogonal bit line (BL) and word line (WL), the resister memory element can be sandwiched between the BL and WL. Therefore, the cell size can be only 4 F^2, which is ideal for ordinary two-dimensional (2D) memory cell design. A problem is that there can be many leak paths between the selected BL and WL. One of the leak paths is shown with a dashed line in Fig. 3.3. Leak paths make writing and reading difficult. To break the leak path, one diode must be used for one resister memory element. Due to the difficulty in fabricating the diode with sufficiently high performance, the cross-point memory has not been commercialized yet. In 2015, an introduction of a cross-point non-volatile memory, i.e., *3D XPoint memory*, for SCM application was announced, although details have not been disclosed [3].

A NAND flash memory cell (Fig. 3.2c) is specially designed to reduce the cell size to 4 F^2. The memory element, i.e., floating gate (FG), is integrated into the transistor, and multiple transistors are serially connected by sharing their sources and drains with the adjacent transistors. This design sacrifices the random accessibility and access speed to smaller cell size.

To further reduce the effective cell size, recent NAND flash memory also uses the multi-bit per-cell storage design and three dimensional (3D) stacking architectures. One memory element of traditional NAND flash memory stores one bit, i.e., 0 or 1. Two different signal levels corresponding to 0 and 1 are used for reading. If the signal difference is large enough and its distribution can be tightly controlled, multi-bit information can be stored into one memory cell. Figure 3.4 shows a case of 2 bits per cell. Its effective cell size is 2 F^2. The multi-level cell (MLC) design has drawbacks in access speed and endurance because writing is done by multiply divided writing to

Fig. 3.3 Cross-point architecture of resistor-based non-volatile memory. Each memory cell requires one diode to break leak path

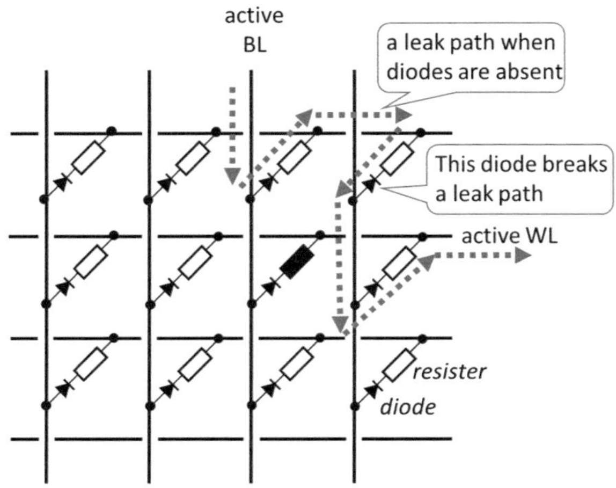

Fig. 3.4 Schematic signal level distribution of multi-bit per-cell NAND flash memory

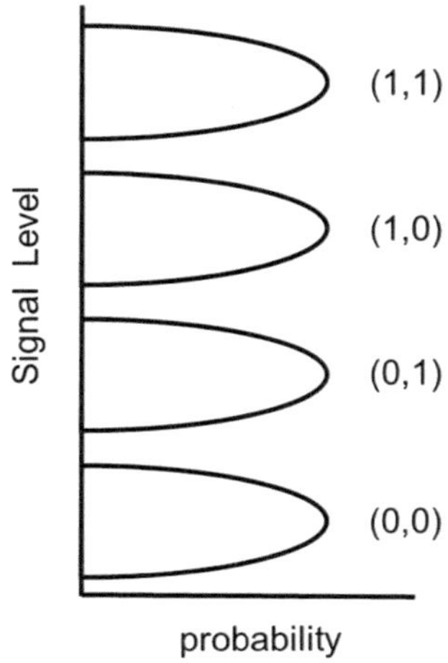

ensure tight distribution. The 3D stacking is a direct way to reduce the effective cell size. A product of 256-Gbit NAND with 3-bit/cell MLC and 48-layer stacking is now available [4]. Although the feature size has not been disclosed, it should be much larger than the most advanced feature size of about 15 nm used for ordinary 2D NAND cells because of the difficulty of 3D stacking.

Access Speed

Access speed determines the memory's positioning in the memory hierarchy. The reading and writing access speeds are not always equal. Abilities of direct overwriting and non-destructive reading are preferred for faster access speed.

The SRAM is a transistor flip-flop logic circuit, and information can be directly over-written and non-destructively read with very fast speed of less than 10 ps based on most advanced CMOS technologies.

The reading process of DRAM is destructive because a charge in the capacitor of the memory cell is discharged for reading. Because DRAM adopts a memory cell array architecture in which multiple memory cells are connected to one WL, all the information of the connected cells are destroyed by one reading. The restoration of the previously stored information of all bit cells connected to the same selected WL is required every time after reading. The writing process of DRAM is also "destructive", because the activation of the selected WL for the writing also destroys the stored information of all connected bit cells. Before writing, the read out of the stored information of all connected bit cells are required for successive restoring [5]. A random access time of commodity DRAMs is typically 40 ns. Note that resistivity-based memory has no such problem of the read-induced bit erase.

The NAND is a slow memory. The electron injection into and electron extraction from the FGs for the respective writing and erasing operations are done by applying high voltages to the memory cell. To increase the memory capacity by reducing effective cell size, typically 32 to 128 transistors are serially connected by connecting their source and drain. Therefore, the writing process becomes complex. The injection and retraction of electrons are done by some assemble units called page and block, respectively. The injection and extraction times are typically 100 μs and 1 ms, respectively. Reading is non-destructive and random access, and is about 10 μs.

Access Power

Access power is the energy required for writing and reading information, i.e., active power.

Static Power

Static power is the energy required to keep information. A SRAM's large static power is due to the leak current between the source and drains and through the gate of field effect transistor (FET). The DRAM requires a cyclic refresh operation to compensate for the leaking charge from the capacitor. Typically, every 10 ms, all stored bits must be read out and restored to avoid information loss. Non-volatile memories do not require power to keep information. However, actual power dissipation depends on the controller of the memory array. The NAND consumes about 10 mW/page when powered.

Random Accessibility

Random accessibility is the ability to randomly access the targeted bit. The SRAM and DRAM have random accessibility. The NAND's writing is by page unit, and its reading is by bit-by-bit.

Non-volatility

Non-volatility is the ability to keep valid data without power supply. The SRAM and DRAM are volatile memories. The NAND is non-volatile.

Retention

Retention is the time during which information is retained. Usually retention of more than 10 years is required. Non-volatile memories that cannot guarantee 10 years of retention may find its applications in cache memory and main memory.

Endurance

Endurance is the guaranteed read/write cycle number without damaging memory. For example, a 1-MHz access cycle requires 3×10^{14} cycles for 10-year operation. This requires memories for moderate-performance computers to have endurance larger than 10^{15}. The SRAM and DRAM satisfy this requirement. The NAND uses charge injection and extraction by applying voltage as high as 12 V through thin oxide film surrounding the FG. This process damages the thin oxide film. The endurance of a simple NAND is limited to about 10^6, and it becomes much smaller for multi-level bit-per-cell NAND.

Temperature

Temperature is the temperature range without information loss. Typical required ranges for commercial and automotive applications are –40 to 85 and –40 to 125 °C, respectively.

3.2 Variety of NVRAM

The characteristics of STT-MRAM, ReRAM, PCRAM, FeRAM, and NOR flash are discussed. Because their development is rapidly advancing, the cited values should be regarded as present typical values.

3.2.1 STT-MRAM

(1) Introduction

The STT-MRAM [6] uses the magnetization direction of ferromagnetic material to store information, the same as HDDs and old magnetic core memory. The essential difference between STT-MRAM and core memory is their magneto-electric information conversion methods for reading and writing. Instead of a classical electromagnetic coil used in core memory, recently developed highly efficient quantum mechanical conversion methods are used in STT-MRAM. This enables an increase in memory capacity while retaining fast access speed and high endurance of magnetic memory and storage.

For information reading, magnetic bubble memory, which was commercialized in the 70s, had already used the anisotropic magneto-resistive (AMR) effect instead of a magnetic coil. The AMR effect is a quantum mechanical effect in which the electrical resistivity of magnetic materials depends on the relative angle between the magnetization direction and current flow direction. The resistivity change in AMR is about 1–2%. After the 80s, giant MR (GMR) [7] and tunnel MR (TMR) [8, 9] effects were discovered in layered structures composed of nano-meter-thick magnetic and non-magnetic films. At present, the practical TMR ratio is as high as 200–300% [10]. The MRAM is a non-volatile magnetic memory that uses the TMR effect for information reading. A commercial product of 16-Mbit MRAM is now available [11].

The MRAM still uses electromagnetic coils for writing information. The coils have no scalability, and its conversion efficiency rapidly degrades with reduced cell size. An MRAM bit cell smaller than 100 nm is not practical. A new method for changing the magnetization direction without using a coil was theoretically predicted in the 90s [12, 13] and experimentally demonstrated in the early 2000s [14–16]. This method is the STT magnetization reversal method, which is scalable. The STT-MRAM uses STT for information writing and TMR for reading.

The fast access speed and high endurance of STT-MRAM are the most important advantages for main memory, cache memory, and non-volatile logic circuit applications.

(2) Principle of Operation

A two-terminal magnetic tunnel junction (MTJ) (Fig. 3.5) made of very thin, typically 1-nm-thick, insulating MgO tunnel barrier sandwiched between two ferromagnetic films is fabricated in the back-end of the line (BEOL) process on a MOS transistor-selecting device (Fig. 3.6). The magnetization direction of one ferromagnetic film (reference layer) is firmly fixed, and that of the other ferromagnetic film (free layer) can have one of two stable directions, which represents the information.

The information is non-destructively read out using the TMR effect of the MTJ. The resistivity is high or low depending if the magnetization directions of the two ferromagnetic films are anti-parallel or parallel, respectively. The TMR ratio is defined as the ratio of the resistivity difference against the lower resistivity. Currently, the typical TMR ratio is 100–200%, which corresponds to double to triple change of the resistivity.

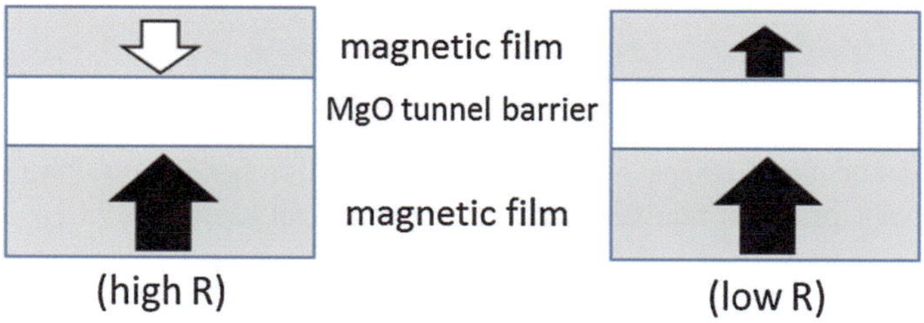

Fig. 3.5 Magnetic tunnel junction (MTJ) for MRAM and STT-MRAM. MTJ is made of about 1-nm-thick tunnel barrier sandwiched between two ferromagnetic electrodes. Anti-parallel and parallel alignments of magnetizations of two electrodes result in higher and lower resistivity, respectively

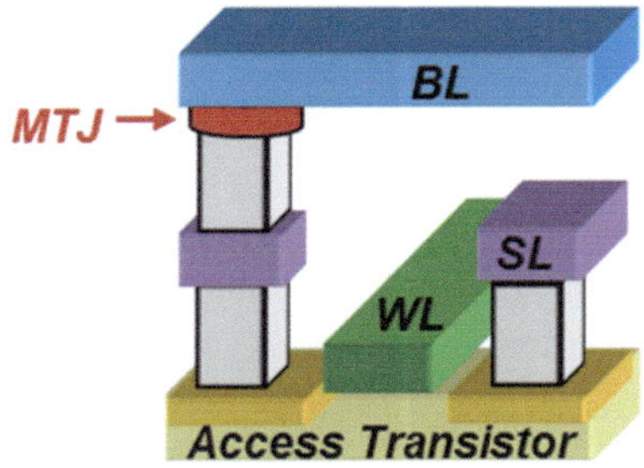

Fig. 3.6 Bit cell of STT-MRAM. Circular MTJ of diameter F can be used allowing ultimate cell size of 6 F^2

Electrical current in magnetic materials has a very different nature from that in non-magnetic materials. The current flow from magnetic materials contains unbalanced numbers of up-spin and down-spin electrons. By flowing this spin-polarized current between the two ferromagnetic metal electrodes of an MTJ, the magnetization direction of the free layer can be reversed depending on the direction of the current flow. The STT-MRAM is a non-volatile memory that operates in a bipolar mode.

The use of perpendicular ferromagnetic films, in which the magnetization direction is perpendicular to the film plane, became standard after 2008 to increase retention and reduce writing power [17].

(3) Characteristic Features

Scalability

Because the bit cell size can be as small as 6 F^2 and F can be smaller than 20 nm, a Gbit-order capacity is expected if the required current for STT-writing can be delivered by such a small transistor. Lowering the STT-writing current is important to enhance memory density. In 2014, a 64-Mbit sample product was announced with F = 90 nm [18]. Currently, MTJs around 30 nm are being intensively studied and material development toward 1× nm is under way. One-Gb STT-MRAM with 22 F^2 and $F = 28$ nm has been reported [19]. Small resistivity change in MTJ makes multi-bit per cell and 3D stacking difficult.

Access Speed

The reading speed depends on the TMR ratio and resistivity of an MTJ. The writing speed is fundamentally limited by the ferromagnetic resonance frequency of materials, and can be about 100 ps. Due to a trade-off relation between writing speed and writing energy, practical read and write speed is several nsec.

Access Power

The most power-consuming process is the magnetization reversal for writing. The lower writing power and longer retention are in a trade-off relation. The introduction of perpendicular ferromagnetic materials [17] was decisive for decreasing writing power while maintaining sufficient retention. The writing voltage is as low as 0.6 V, and the writing current density is below 1 MA/cm^2. Writing with 0.09 pJ with 2 ns, 50 μA, and 0.6 V has been reported [20, 21].

Static Power

No power is consumed to keep information.

Random Accessibility

OK

Non-volatility

OK

Retention

More than 10 years is guaranteed for 30-nm size bits. Although the retention can degrade along with bit-cell shrinkage, there is no fundamental limit for ferromagnetic materials to guarantee 10 years with 1 × nm size.

Endurance

The tunnel barrier can be damaged if the tunneling current density is too high. An MRAM's endurance is higher than 10^{15}. The recent rapid decrease in STT current to below 1 MA/cm^2 seems to make more than 10^{15} possible, also for STT-MRAM. The STT-MRAM has the best endurance among non-volatile memories.

Temperature

In general, magnetic devices are robust against radiation and temperature. The MRAM is expected to work up to 140 °C.

3.2.2 ReRAM

(1) Introduction

The ReRAM uses an electrical induced creation (SET) or rupture (RESET) of nano-size conductive filament inside an insulating film. The presence and absence of filament results in lower and higher resistivity, respectively. This RAM is also called RRAM or memristor [22].

The phenomenon of voltage-induced resistivity change of insulators had been known since the 60s. Around 2000, when apprehension regarding the performance limit of NAND was emerging, memory operation of ReRAM was demonstrated using $Pr_{0.7}Ca_{0.3}MnO_3$ thin film [23], which triggered intensive studies of ReRAM with a variety of materials.

Because of possible ultra-high density and much faster access speed than NAND, ReRAM is a candidate for SCM and post-NAND flash memory. The ReRAM shares many characteristics with PCRAM. A 64-kB ReRAM with a 180-nm process was embedded in commercial 8-bit microcontrollers in 2013 [24].

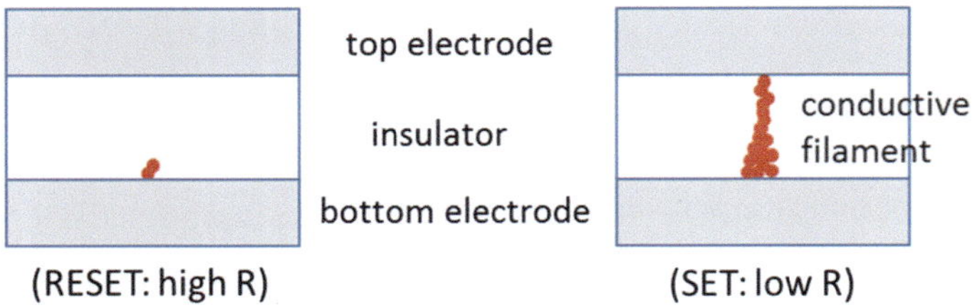

Fig. 3.7 Memory element made of typically 10-nm-thick insulating film sandwiched between two electrodes is used for ReRAM. Conductive filament is formed or ruptured in insulating film by applying voltage to electrodes for writing. Memory elements are used in 1T-1R-bit cell structure or cross point cell structure to form bit cells

(2) Principle of Operation

A two-terminal memory cell made of a typically 5–10 nm-thick insulating film sandwiched between two electrodes is fabricated using the BEOL process on a selector device, i.e., MOS transistor or diode. There are two types of ReRAM, that which uses conductive filament made of oxygen vacancy in the metal-oxide insulator, called OxRAM, and conductive bride RAM (CBRAM), which uses metal bridge filament inside a solid electrolyte (Fig. 3.7).

The OxRAMs use metal-oxide materials, such as HfO_2, Ta_2O_5, NiO, ZrO_2, and TiO_2, which are already familiar in the Si CMOS LSI process. When the compositions of these materials are stoichiometric, they are highly electrically insulating. They become conductive when oxygen vacancies are introduced. The oxygen vacancy forms a conductive filament with a diameter of about 10 nm. The electrical-field-induced movement of oxygen atoms is the main mechanism for the introduction of oxygen vacancy. Diffusion of oxygen atoms by local Joule heating is also considered to contribute to the vacancy formation. Depending on the combination of materials for metal-oxide insulator film and metallic electrodes, the writing voltage can be either bipolar or unipolar. Because unipolar ReRAM can use a diode for its selecting device, a cross point memory structure, which is suitable for higher capacity, can be adopted. However, due to the constraint of switching speed and stable operation, bipolar ReRAM has been a focus of recent developments. Information is read by measuring the 10–100 times difference in resistivity.

The CBRAM-type ReRAM uses electrochemical reaction to form the conductive filament made of metal, such as Ag and Cu, in a solid electrolyte such as GeS_2, AgSe, and Ta_2O. Local Joule heating also contributes to the rupture of the filament. The CBRAM is slower and exhibits less endurance than OxRAM. However, its resistivity change ratio is as large as 5–6 orders of magnitude. Therefore, CBRAM can work as a good electrical switch suitable for the configuration memory element of field-programmable-gate-arrays (FPGAs).

(3) Characteristic Features

Scalability

The large resistivity change ratio enables multi-bit per cell and 3D stacking. The cell area can be 4 F^2 and be as small as the size of a conducting filament of about 10 nm, in principle. However, in practice, the present typical writing current 50 μA is too large for the selecting device. Furthermore, the writing current does not scale down with cell size because the filament size is much smaller than the cell size. Materials with smaller writing current should be developed. A 32-Gb OxRAM with $F = 24$ nm [25] and a 16-Gb CBRAM with $F = 27$ nm [26] have been reported.

Access Speed

The OxRAM can be as fast as several ns. Practical reported speed is several ns to 100 ns. The CBRAM works typically with a speed of 100 ns to 100 μs.

Access Power

RESET is the most power-consuming process. Typically, around a 50 μA current with 1–3 V pulses is used. The reported switching energy is typically 0.1–1000 pJ. Because of the nano filamentary nature of the operating principle, the writing power cannot be scaled down along with memory cell size. Therefore, material development is required.

Static Power

No static power is consumed.

Random Accessibility

OK

Non-volatility

OK

Retention

The amount of oxygen vacancy in the conductive filament can be reduced by the movement of oxygen atoms induced by the electrical field applied for the reading or environmental temperature, resulting in an increase in the resistivity of the SET state. Smaller writing current induces a smaller amount of the oxygen vacancy resulting in shorter retention. Ten-year retentions at 85 °C has been reported for a commercial product [24]. Ten years at 125 °C is a target of future development.

Endurance

Repetition of reading and writing can cause an insufficient rupture of the filament, resulting in disappearance of resistivity change. Retention over 10^5 cycles was guaranteed in a recent commercial product [24]. More than 10^9 cycles is expected.

Temperature

A commercial product guarantees –40 to 85 °C [24].

3.2.3 PCRAM

(1) Introduction

Some compounds (chalcogenides) containing Se or Te can have two material phases, i.e., amorphous and poly-crystalline. By controlling the heating and cooling processes, one of the two phases can be selectively stabilized at room temperature. A large difference in the resistivity between the two phases is used for PCRAM, which is also called PRAM or PCM.

Just after a report on the electrical control of the material phases in 1968 [27], a PCRAM was commercialized but failed to survive due to it large writing current. The two material phases also have large difference in optical reflectivity. By controlling the reflectivity change with laser heating, a rewritable optical disk was commercialized. A chalcogenide material $Ge_2Sb_2Te_5$ (GST) [28], which was developed for DVD-RAM, is now widely used for PCRAM.

After 2000, interest in PCRAM was renewed due to apprehension regarding NAND scalability and the desire for new non-volatile memory with faster access speed and higher endurance than NAND. A problem of early PCRAM, i.e., large writing current, was expected to be overcome by reducing the memory cell size to well below 1 μm.

The PCRAM shares many characteristics with ReRAM. The SCM and replacement of NOR flash are targeted applications for PCRAM. The PCRAM was commercialized the earliest among new non-volatile memories. The 512-Mbit PCRAM is now in mass production [29].

(2) Principle of Operation

A two-terminal memory cell composed of a chalcogenide thin film with a heater and two electrodes is fabricated in the BEOL process on a selecting device of a MOS transistor or diode (Fig. 3.8).

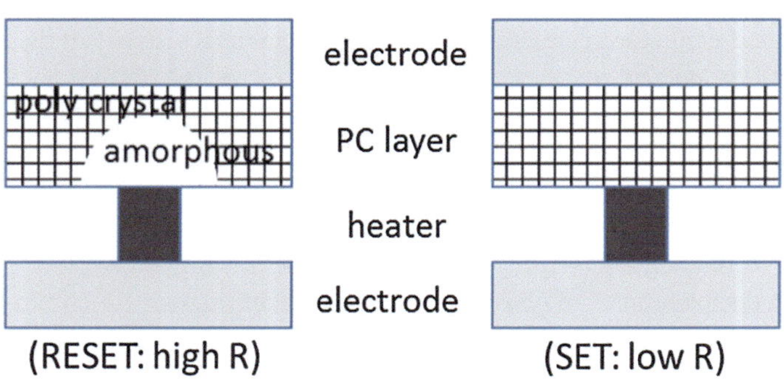

Fig. 3.8 Memory element of PCRAM. Phase change (PC) material, GST, is heated above melting point or crystallization temperature by current flow through heater. Controlling cooling process results in either amorphous phase with higher resistivity or crystal phase with lower resistivity. Memory elements are used in 1T-1R bit cell structure or cross point cell structure to form bit cells

Joule heating by flowing current through the heater increases the temperature of the contacting part of the phase change material above the melting point or crystallization temperature. Rapid cooling from above the melting point results in the amorphous phase (RESET). Relatively slow cooling from above the crystallization temperature results in the poly-crystal phase (SET). The GST has a low melting point of 620 °C and a low crystallization temperature of 160 °C, which are useful to reduce writing power.

Information reading uses a resistivity difference as large as 2 orders of magnitude between the two phases. The randomly arranged atomic structure of the amorphous phase scatters the conducting electrons much more than a well aligned atomic structure of the crystal phase and results in a larger electric resistivity. The PCRAM can work with unipolar voltage, which has suitably higher capacity.

(3) Characteristic Features

Scalability

Due to its large resistivity difference ratio, a multi-bit-per-cell structure and 3D stacking are possible. In principle, cell size can be as small as 4 F^2. However, in practice, large RESET current is a problem. A large selecting device is required to supply large current and limits density. Thermal design to prevent the possible heating effect of local temperature higher than 600 °C for RESET to the adjacent bit is also required for high capacity. A fabrication of 8-Gbit PCRAM has been reported [30].

Access Speed

Irrespective of the previously stored bit, new information can be directly written. The writing speed is determined by the speed of the phase change. The SET is the speed-limiting process, which can be as fast as 10 ns if a phase change material with low crystallization temperature is used. However, faster SET speed and longer retention is in a trade-off relation. In practice, about 1-μs operation is typical.

A memory cell shows large resistivity change, as large as two orders of magnitude. Distribution limits the practical available resistivity change to about one order for 1-Gbit integration. The reading speed is about 50–100 ns [31].

Access Power

For RESET, large current to heat the phase change material to above melting point is needed. A typical writing current pulse is about 100 μA and 10 ns long. This heating process makes the access power of PCRAM relatively larger than those of other non-volatile memories. For reducing in the writing current, new device structures that can limit the volume of the phase changing area or new phase change materials are expected.

Static Power

No static power is consumed.

Random Accessibility

OK

Non-volatility

OK

Retention

Because the crystallization temperature of phase change materials is designed to be low, the amorphous state can de-stabilize due to the surrounding temperature. At 85 °C, a retention of 15 months was reported [32]. Retention was reported to be improved by using an appropriate dielectric capping layer [33].

Endurance

As writing repeats the phase change, the atomic arrangement can deviate from the pure crystalline state or amorphous state. Local compositional segregation of phase change material and chemical reaction among phase change material, heater, and electrodes also causes reduction in resistivity change and snapping of memory cells. Typical endurance is around 10^6–10^8 [31].

Temperature

Because of the low crystallization temperature, PCRAM may not be suitable for automobile and industrial applications.

3.2.4 FeRAM

(1) Introduction

The FeRAM is a nonvolatile memory in which a capacitor part of the DRAM cell is replaced with ferroelectric thin films (ferroelectric capacitor) such as $Pb(Zr,Ti)O_3$ (PZT) or $SrBi_2Ta_2O_9$ (SBT).

Full-scale mass production began in about 2000, and several companies mass-produce FeRAM now. The memory capacity is low, but FeRAM is a non-volatile memory that is suitable for high-speed rewritable applications. It is used for IC cards, security chips to prevent forgery, event recorders, and smart meters. Its adoption began in the area of bridging the access speed gap between DRAM and SRAM and NAND flash memory including the page buffer of SSDs and other low-speed non-volatile memory. The specifications of commercial FeRAM are 4-Mb capacity, 10^{14}-times endurance, and access speed of 55 ns. Although it is slow, the international technology roadmap for semiconductors (ITRS) expects steady progress in comparison with other memories.

(2) Principle of Operation

A ferroelectric substance is a ceramic material used for piezoelectric elements. The electric charge remains even if the voltage applied to the ferroelectric substance returns to zero after the voltage application, as shown in Fig. 3.9. The electric charge

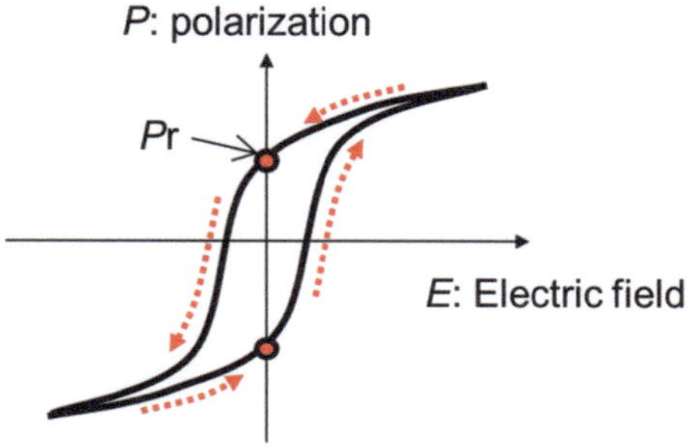

Fig. 3.9 Hysteresis loop of ferroelectric substance

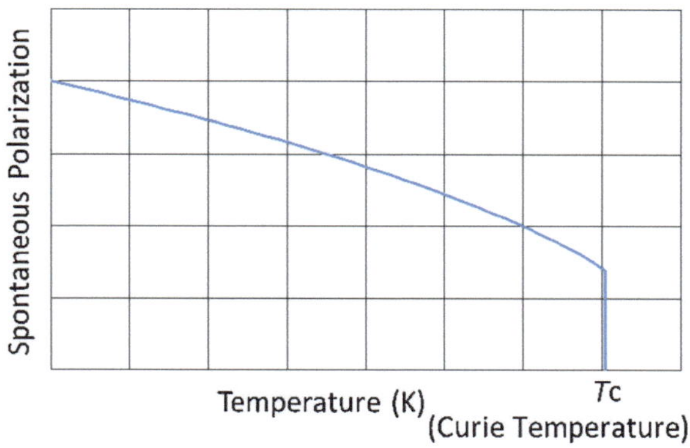

Fig. 3.10 Curie temperature of ferroelectric substance

is called a remanent polarization or spontaneous polarization. The representative ferroelectric material used in FeRAM are PZT and SBT. The ferroelectric substance has a Curie temperature (approximately 400 °C in PZT) and ferroelectricity disappears at higher temperature (Fig. 3.10).

There are two types of FeRAM, capacitor, and FET. The FET type has a ferroelectric layer in its gate stack. The surface conductivity is modulated using the remanent polarization of the ferroelectric layer, as shown in Fig. 3.11. This type of memory has never been in production because of its short data retention, difficulty in miniaturization, and high write voltage. However, reports have recently been published regarding longer data retention of more than 3.5 months [34] and scaling potential by using ferroelectric-doped HfO_2 [35].

Capacitor-type ferroelectric substance memory (1T-1C type) plays a key role in the development of ferroelectric memory. The 1T-1C-type exhibits non-volatility by using a ferroelectric substance for a capacitor of DRAM. Figure 3.12 shows the structure of a 1T-1C-type memory cell. A ferroelectric capacitor is formed in the upper part of the general MOSFET through an inter-layer insulation film and connected to the source of the FET. The circuit of the memory cell is shown in Fig. 3.13. By the polarization state of the ferroelectric film, information of one bit distinguishes "1" or "0". The direction of the remanent polarization is determined by the polarity of the voltage applied to the ferroelectric capacitor.

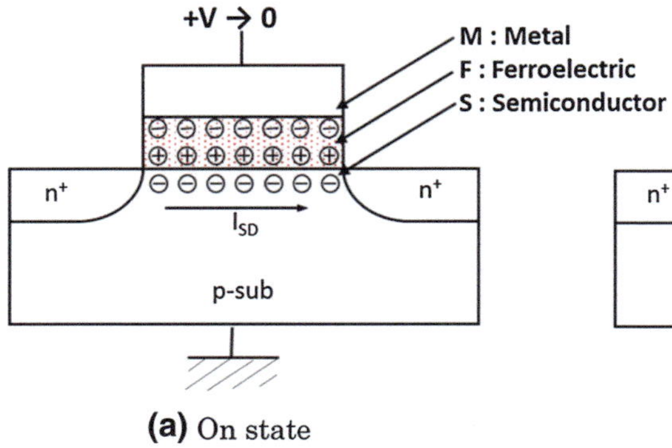

Fig. 3.11 Operation principle of FET-type FeRAMs

Fig. 3.12 Structure of 1T-1C-type memory cell

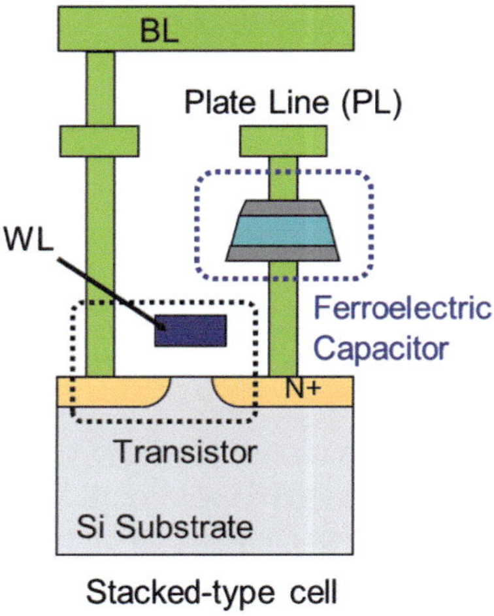

Fig. 3.13 Circuit of 1T-1C-type memory cell

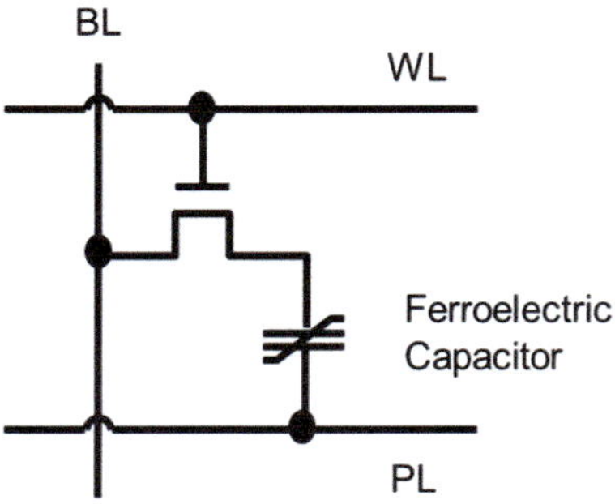

The write access to a memory cell is done in the following procedure. First, voltage is applied to the WL to choose and turn on a select transistor. Then, a pulse {0→high level→0 } is applied to the plate line (PL) after applying the voltage to the BL (1 for high voltage, 0 for low voltage). Data are recorded in a ferroelectric substance for the remanent polarized charge by the voltages {BL: high level, PL: 0} being applied for a ferroelectric capacitor at the time of 1, and those {BL: 0, PL: high level} being applied at the time of 0.

Reading is done by applying voltage to a ferroelectric capacitor and taking out an electric charge. By applying a pulse to the PL, the voltage is generated on the BL and the voltage depends on the residual polarization. Like DRAM, data are retrieved using a sense amplifier by amplifying the small signal voltage corresponding to 1 or 0. The polarization state of the ferroelectric capacitor is reset to one direction by the reading action. Therefore, it is necessary to write after the reading again (destruction reading).

(3) Characteristic Features

Scalability

The memory size and capacity of a mass-production FeRAM are 0.71 μm^2 and 4-Mb. In a paper, 128-Mb FeRAM was reported [36]. The F factor is about 23 times in current production due to difficulties in etching of ferroelectric capacitors and preventing process degradation during the wiring and passivation process. To proceed with the miniaturization of FeRAM, breakthrough in etching technology and development of 3D capacitor fabrication are key. For memory product specification, improvements in low-voltage operation and higher storage temperature are required.

Access Speed

The access speed is 55 ns with a mass-production product and 30–40 ns with a development product [37]. The fact that FeRAM is slower than SRAM and DRAM is one of the problems with practical applications. The reason of the slow access speed is due to the time required to charge a large ferroelectric capacitor, not to the intrinsic property of the material. The write time can be shortened by reducing the capacitor volume, but this results in degradation of data retention. This means that the write time can be shortened if the application does not require much longer retention.

Access Power

When one bit is read and written back again, the action point in the voltage-polarization plot goes around a hysteresis loop, as shown in Fig. 3.9. It uses energy of $2\ \mathrm{Pr} \cdot \mathrm{S} \cdot 2\ \mathrm{Vc} = 0.036$ pJ in a typical case of coercive voltage $\mathrm{Vc} = 0.5$ V, $2\ \mathrm{Pr} = 8.5\ \mu\mathrm{C/cm}$, and $\mathrm{S} = 0.423\ \mu\mathrm{m}^2$ (ITRS2015). The access power becomes 0.036 pJ/100 ns $\cdot$ 8 bits = 2.9 μW for 1 byte (8 bits) at 100 ns.

Static Power

There is no leak pass in a memory cell of FeRAM. In a data maintenance state, no voltage is applied to the WL, BL, PL. Therefore, including the gate leak of the select transistor, the electricity consumption of the memory cell is almost zero.

The actual electricity performance of a FeRAM chip depends on the designs of the power supply and interface circuits resulting in standby current of about 10 μA (at 3 V 30 μW) and dynamic current of about 1 mA (at 3 V 30 mW). This large power consumption originates from the WL driver, PL driver, peripheral circuits such as sense amplifiers, and the power supply circuit including a charge pump or regulator.

Random Accessibility

The FeRAM has no restriction for random access.

Non-volatility

The FeRAM has non-volatility in the use of specifications.

Retention

Two modes of data retention degradation are known. One is relaxation, in which the spontaneous polarization in the maintaining state becomes small, and the other is imprint, in which writing of the opposite polarity becomes difficult (Fig. 3.14). These two retention degradation modes progress with a logarithm of time (change decreases with progress of time in Fig. 3.15). Generally, high temperature tends to accelerate degradation, but there is a product guaranteeing a memory access and data retention to 125 °C [38].

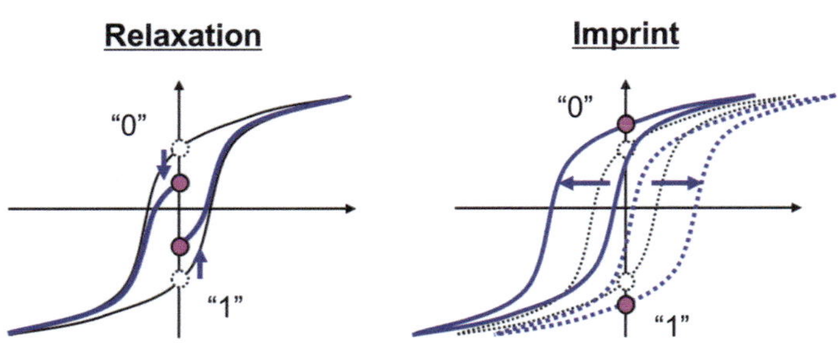

Fig. 3.14 Modes of data retention degradation; relaxation and imprint

Fig. 3.15 Logarithm plot regarding data retention

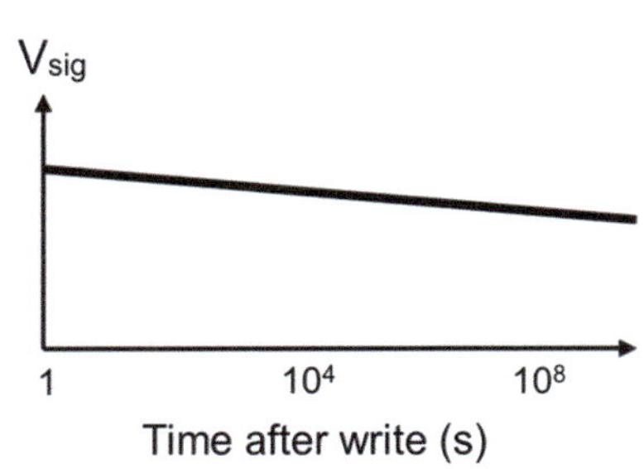

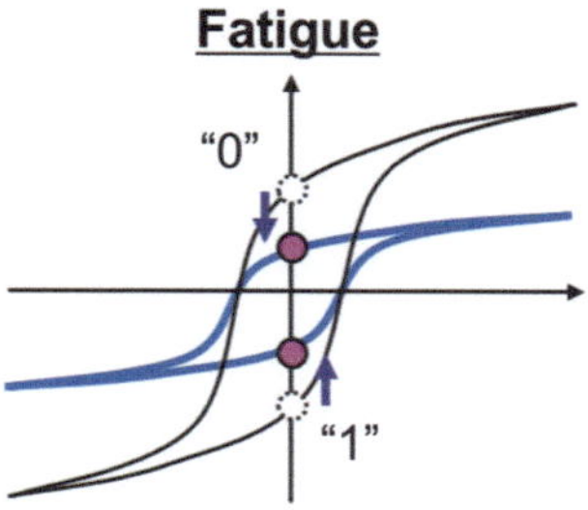

Fig. 3.16 Mode of endurance

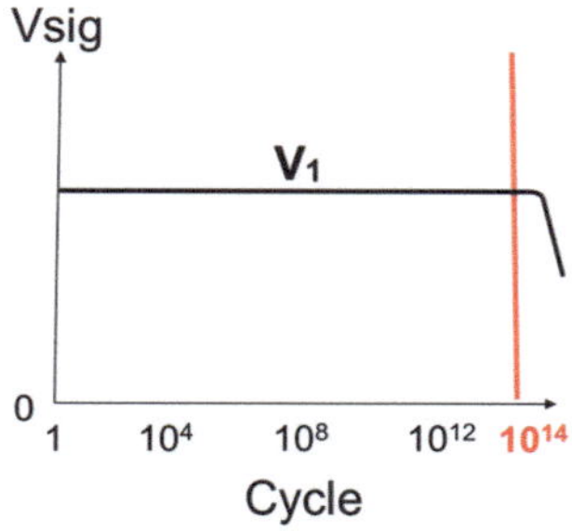

Fig. 3.17 Logarithm plot regarding endurance

Endurance

Endurance has a phenomenon called fatigue in which attainable ferroelectric polarization decreases by repeating the reading and writing, as shown in Figs. 3.16 and 3.17. Fatigue is known to accelerate by applying an electric field and temperature. There is currently a product to guarantee 10^{14} read/write cycles [37]. It has been shown from an acceleration examination that the ferroelectric capacitor can have a rewritable life of more than 10^{15} times with recent ferroelectric manufacture technology, electrode technology, and progress in the processing technique. Rewritable life is determined from the evaluation time and limit of the guarantee technology. Other non-volatilization memory faces the same problem.

Temperature

The higher temperature limit for operation and data storage is 125 °C. It is limited by transistor drivability and leakage and the retention of ferroelectric materials.

The lower temperature limit is about –40 °C. It is limited by increasing the coercive voltage with decreasing temperature, which results in increased operating voltage.

3.2.5 NOR Flash

(1) Introduction

Flash memory has been developed as bulk electrically erasable programmable read-only memory (EEPROM). The feature of flash memory is that data are stored by

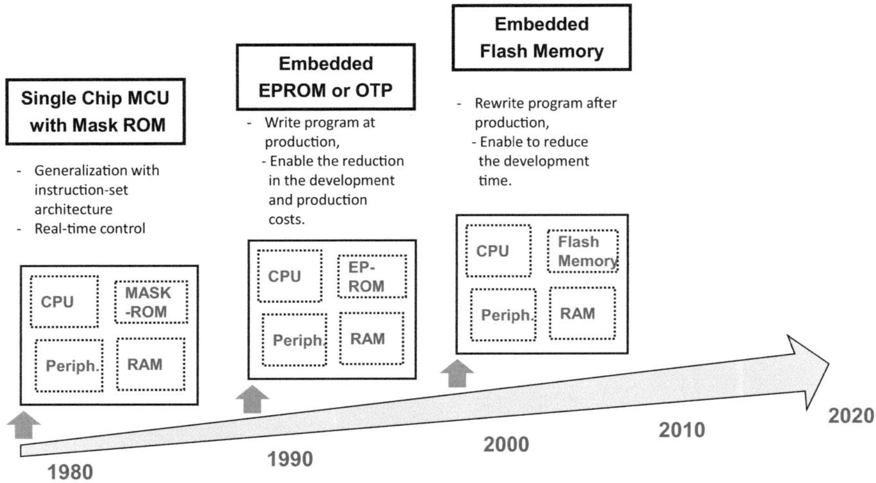

Fig. 3.18 Trend in microcontroller with embedded non-volatile memory

"Physics". In contrast, data in SRAM are stored by "Circuits" and data in DRAM are stored by "Electrons".

From the end of the 70s through the 80s, single-chip microcontrollers with embedded mask ROM appeared, and it became possible to make generalization with the instruction-set architecture and real-time control. Therefore, large advances have been made in terms of performance and ease of use. In the late 80s, microcontrollers with erasable programmable read only memory (EPROM) or one time programmable read only memory (OTP) appeared, and it became possible for users to write program data at the production stage. Therefore, the development and production costs greatly decreased in the last half of the 90s, microcontrollers with embedded flash memory (flash-MCU) appeared, and it became possible to rewrite program data after production. Therefore, a mass production setup became possible at program development completion, and the development period shortened. In addition, a large change occurred regarding production and distribution costs with the generalization of microcontrollers. Currently, Flash-MCU is the mainstream for microcontrollers (Fig. 3.18).

In this chapter, NOR flash memory, which is the main of flash memory installed in microcomputers, is described.

(2) Principle of Operation

(2.1) Operating Principle of Flash Memory

A floating-gate-type flash cell has a MOS gate composed of two layers. The lower layer sandwiched between insulating layers is called an FG, which can store electrons. The negative charge stored in the FG increases the MOS threshold voltage (V_{th}) and decreases the current flow from drain to source. The presence and absence of the negative charge in the FG corresponds to "0" and "1", respectively (Fig. 3.19).

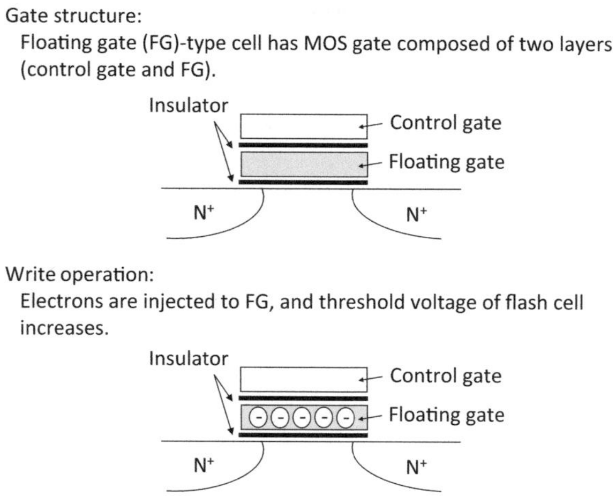

Fig. 3.19 Gate structure and write operation mechanism of floating-gate-type flash cell

(2.2) Writing and Erasing Method of Flash Memory

To cross the barrier of the silicon oxide film, energy of about 3.2 eV is necessary. Therefore, it is necessary to increase the electron energy using various methods. One method is to accelerate electrons with a voltage between the source and drain to create channel hot electrons. Another is called Fowler-Nordheim (FN) tunneling, in which electrons are extracted by applying a forced voltage (or pushed) between the gate and source or well. Examples described here is a typical scheme. The channel hot electron injection method is used for write operation, and FN tunneling is used for erase operation (Fig. 3.20).

(2.3) Types of Flash Memory

There are two types of flash memory, NOR and NAND (Fig. 3.21).

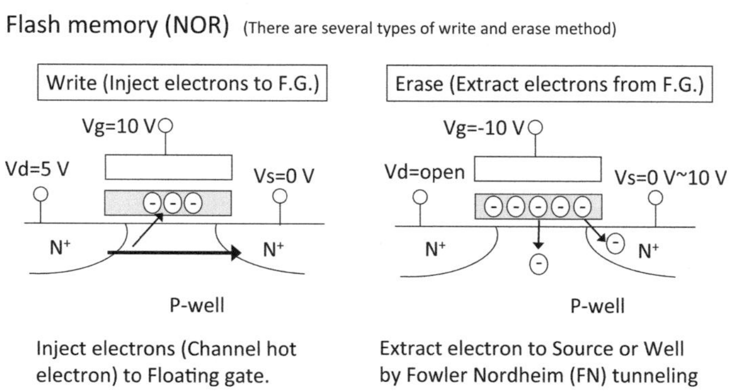

Fig. 3.20 Write and erase operation of NOR-type flash memory

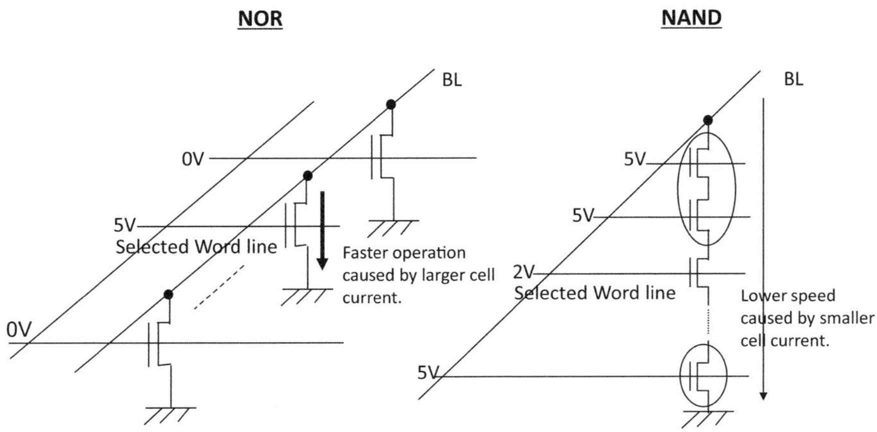

Fig. 3.21 Comparison of NOR- and NAND-type flash memory

In NOR flash memory, all bit cells are directly connected to the BL made of low-resistance metal. This results in a faster operation of NOR flash memory. On the other hand, the metal is not suitable for fine processing, compared with Si. Therefore, the cell size of NOR flash inevitably becomes large resulting in small memory capacity.

In NAND flash memory, several bit cells are connected in series to form a unit, and the connection to the metal BL is done by the unit. Therefore, cells with a smaller pitch than the processing pitch of the metal can be configured, and large memory capacity is possible. However, the serial connection of a large number of cells causes large resistance, which slows the access speed.

(3) Characteristic Features

The characteristic features of flash memory, which is embedded in MCU, are described below.

Scalability

Current flash memory used for an embedded program memory in MCU has a capacity of ~32 Mb with a cell size of ~30 F^2. Currently, 28-nm flash memory is under development [39]. Its capacity is ~128 Mb and cell size is 60–70 F^2.

Access Speed

Read access time is less than 10 ns.

Writing and erasing time of flash memory are slower. In a usual case, page or block size of data is written/erased at the same time to prevent slow operation.

Access Power

Access power is less than 0.1 μJ at write access.

To cross the barrier of the silicon oxide film, electrons need to have energy of about 3.2 eV. Therefore, it is necessary to increase the electron energy in various ways.

Static Power

Zero.

Random Accessibility

Yes

Non-volatility

Yes

Retention

10 years.

If there is a defect in the gate oxide film, all the electrons in the FG will be lost. The defect may become worse with the writing/erasing cycle.

Endurance

$>10^5$ cycles.

If there is a defect in the gate oxide film, all the electrons in the FG will be lost. The defect may become worse with the writing/erasing cycle.

Temperature

The higher temperature limit for operation and data storage is 105°C.

If there is a defect in the gate oxide film, all the electrons in the FG will be lost. The defect may become worse with the writing/erasing cycle.

3.3 Summary

No matter which non-volatile memory will be commercialized, it cannot be a universal memory for every application. Combination of various volatile and non-volatile memories by considering their advantages and disadvantages is a realistic solution. Understanding the characters of each non-volatile memory is also important (Figs. 3.22, 3.23 and Table 3.2).

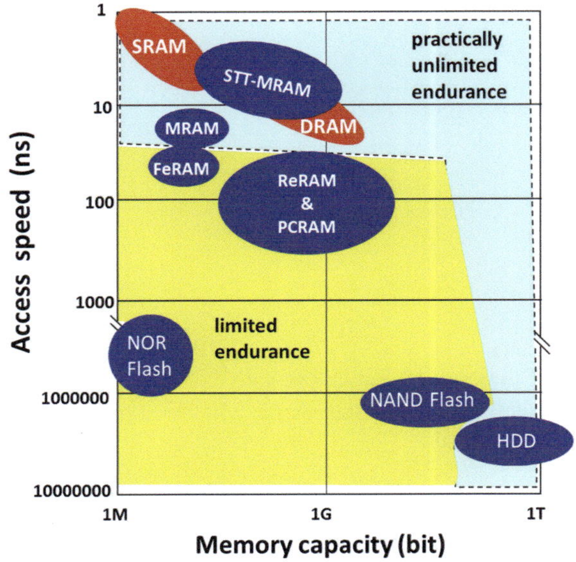

Fig. 3.22 Typical positioning of memory and storage devices

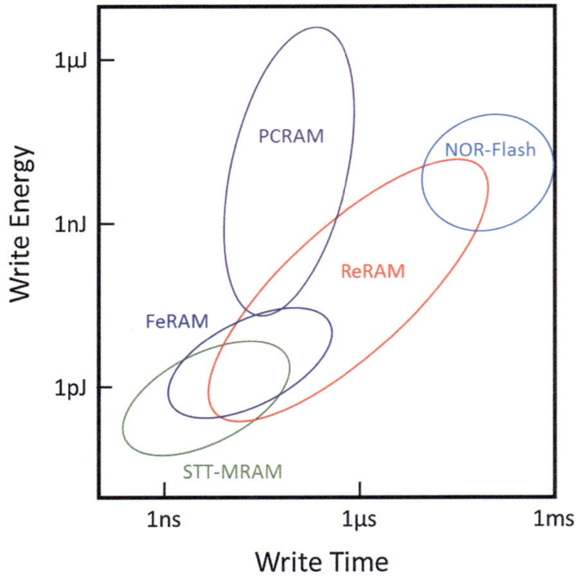

Fig. 3.23 Typical write energy and write time of non-volatile memories. Plots for STT-MRAM, ReRAM, and PCRAM are drawn based on accumulated data of Ref. [40]. Plot for NOR type flash is based on typical value of current products. Plot for FeRAM are based on FEP summary of Ref. [41]

Table 3.2 Typical characteristics of emerging non-volatile memories. (e), (d) and (p) correspond to expected, demonstrated, and product values, respectively

	STT-MRAM	ReRAM	PCRAM	FeRAM	NOR flash
Scalability	(e) 6 F^2	(e) 4 F^2, 3D	(e) 4 F^2, 3D	(e) 20 F^2, 2D	(e) ~30 F^2
	(d) 1 Gb@28 nm & 22 F^2	(d) 32 Gb	(d) 8 Gb@20 nm	(d) 128 Mb@130 nm	(d) 128 Mb@28 nm
	(p) 64 Mb@90 nm	(p) 512 kb	(p) 512 Mb@60 nm	(p) 4 Mb@130 nm	(p) 32 Mb@90 nm
Access speed	(d) 2–4 ns	(d) 10 ns–10 us	(d) 20 ns–1 us		(d) 5 ns @Read cycle
				(p) 55 ns	
Access power	(d) 0.09 pJ	(d) 0.1 pJ–1 nJ	(d) 20 pJ–100 nJ		
				(p) 55 ns	
Static power	0	0	0	0	0
Random accessibility	○	○	○	○	X
Non-volatility	○	○	○	○	○
Retention			(e) 10 years <		
			(d) 15 months@85° C		
	(p) 10 years <	(p) 10 years <		(p) 10 years@85° C <	(p) 10 years <
Endurance	(e) 10^{15} <	(e) 10^9 <			
			(d) 10^6 -10^8		
		(p) 10^5 <		(p) 10^{14} <	(p) 10^5 <
Temperature	(e) <140 °C		(e) <85 °C		
		(p) –40 to 85 °C		(p) –55 to 125 °C	(p) –40 to 105 °C

References

1. Dennard, R.H., Gaensslen, F.H., Yu, H.-N., Rideout, V.L., Bassous, E., LeBlanc, A.R.: Design of ion-implanted MOSFET's with very small physical dimensions. IEEE J. of Solid-State Circuits **9**, 256–268 (1974). October
2. Fujita, S., Nomura, K., Noguchi, H., Takeda, S., Abe, K.: Novel nonvolatile memory hierarchies to realize "Normally-Off Mobile Processors". In: ASPDAC2014, Singapore, January 2014
3. Micron: 3D XPoint Technology. https://www.micron.com/about/emerging-technologies/3d-xpoint-technology
4. Samsung: Samsung Electronics Begins Mass Producing Industry First 256-Gigabit, 3D V-NAND Flash Memory. http://www.samsung.com/semiconductor/about-us/news/22659
5. Itoh, K.: VLSI Memory Chip Design. Springer, Berlin (2001). doi:10.1007/978-3-662-04478-0
6. Ando, K., Fujita, S., Ito, J., Yuasa, S., Suzuki, Y., Nakatani, Y., Miyazaki, T., Yoda, H.: J. Appl. Phys. **115**, 172607 (2014)
7. Baibich, M.N., Broto, J.M., Fert, A., Nguyen Van Dau, F., Petroff, F., Etienne, P., Creuset, G., Friederich, A., Chazelas, J.: Phys. Rev. Lett. **61**, 2472 (1988)
8. Miyazaki, T., Tezuka, N.: J. Magn. Magn. Mater. **139**, L231 (1995)
9. Moodera, J.S., Kinder, L.R., Wong, T.M., Meservey, R.: Phys. Rev. Lett. **74**, 3273 (1995)
10. Yuasa, S., Djayaprawira, D.D.: J. Phys. D Appl. Phys. **40**, R337 (2007)
11. Everspin: 16Mb MRAM. https://www.everspin.com/density/16384
12. Slonczewski, J.C.: J. Magn. Magn. Mater. **159**, L1 (1996)
13. Berger, L.: Phys. Rev. B **54**, 9353 (1996)
14. Katine, J.A., Albert, F.J., Buhrman, R.A., Myers, E.B., Ralph, D.C.: Phys. Rev. Lett. **84**, 3149 (2000)
15. Huai, Y., Albert, F., Nguyen, P., Pakala, M., Valet, T.: Appl. Phys. Lett. **84**, 3118 (2004)

16. Kubota, H., Fukushima, A., Ootani, Y., Yuasa, S., Ando, K., Maehara, H., Tsunekawa, K., Djayaprawira, D.D., Watanabe, N., Suzuki, Y.: Jpn. J. Appl. Phys. **44**, L1237 (2005)
17. Kishi, T., Yoda, H., Kai, T., Nagase, T., Kitagawa, E., Yoshikawa, M., Nishiyama, K., Daibou, T., Nagamine, M., Amano, M., Takahashi, S., Nakayama, M., Shimomura, N., Aikawa, H., Ikegawa, S., Yuasa, S., Yakushiji, K., Kubota, H., Fukushima, A., Oogane, M., Miyazaki, T., Ando, K.: IEEE International Electron Devices Meeting , pp. 309–312 (2008)
18. Everspin: 64 Mb Spin-Torque MRAM—DDR3 DRAM Compatioble. https://www.everspin.com/64mb-spin-torque-mram-ddr3-dram-compatible
19. Park, C., Kan, J.J., Ching, C., Ahn, J., Xue, L., Wang, R., Kontos, A., Liang, S., Bangar, M., Chen, H., Hassan, S., Gottwald, M., Zhu, X., Pakala, M., Kang, S.H., Int, I.E.E.E.: Electron Devices Meet. **26**(2), 1–4 (2015)
20. Kitagawa, E., Fujita, S., Nomura, K., Noguchi, H., Abe, K., Ikegami, K., Daibou, T., Kato, Y., Kamata, C., Kashiwada, S., Shimomura, N., Ito, J., Yoda, H.: IEEE International Electron Devices Meeting, pp. 26.2.1–26.2.4 (2012)
21. Saida, D., Shimomura, N., Kitagawa, E., Kamata, C., Yakabe, M., Ohsawa, Y., Fujita, S., Ito, J.: Intermag 2014 EC-5 (2014)
22. Strukov, D.B., Snider, G.S., Stewart, D.R., Williams, R.S.: Nature **453**, 80 (2008)
23. Liu, S.Q., Wu, N.J., Igunatiev, A.: Appl. Phys. Lett. **76**, 2749 (2000)
24. Panasonic: Panasonic starts world's first mass production of ReRAM Mounted Microcomputers. http://news.panasonic.com/press/news/official.data/data.dir/2013/07/en130730-2/en130730-2.html
25. Liu, T.-Y., Yan, T.H., Scheuerlein, R., Chen, Y., Lee, J.K.Y., Balakrishnan, G., Yee, G., Zhang, H., Yap, A., Ouyang, J., Sasaki, T., Addepalli, S., Al-Shamma, A., Chen, C.-Y., Gupta, M., Hilton, G., Joshi, S., Kathuria, A., Lai, V., Masiwal, D., Matsumoto, M., Nigam, A., Pai, A., Pakhale, J., Siau, C.H., Wu, X., Yin, R., Peng, L., Kang, J.Y., Huynh, S., Wang, H., Nagel, N., Tanaka, Y., Higashitani, M., Minvielle, T., Gorla, C., Tsukamoto, T., Yamaguchi, T., Okajima, M., Okamura, T., Takase, S., Hara, T., Inoue, H., Fasoli, L., Mofidi, M., Shrivastava, R., Quader, K.: IEEE International Solid-State Circuits Conference (ISSCC) Digest of Technical Paper 2013, pp. 210–211 (2013)
26. Fackenthal, R., Kitagawa, M., Otsuka, W., Prall, K., Mills, D., Tsutsui, K., Javanifard, J., Tedrow, K., Tsushima, T., Shibahara, Y., Hush, G.: IEEE International Solid- State Circuits Conference (ISSCC) Digest of Technical Papers, pp. 338–339 (2014)
27. Ovshinsky, S.R.: Phys. Rev. Lett. **21**, 1450 (1968)
28. Yamada, N., Ohno, E., Nishiuchi, K., Akahira, N., Takao, M.: J. Appl. Phys. **69**, 2849 (1991)
29. Samsung: Samsung announces production start-up of its next-generation nonvolatile memory PRAM. http://www.samsung.com/semiconductor/insights/news/4097
30. Choi, Y., Song, I., Park, M.-H., Chung, H., Chang, S., Cho, B., Kim, J., Oh, Y., Kwon, D., Sunwoo, J., Shin, J., Rho, Y., Lee, C., Kang, M.G., Lee, J., Kwon, Y., Kim, S., Kim, J., Lee, Y.-J., Wang, Q., Cha, S., Ahn, S., Horii, H., Lee, J., Kim, K., Joo, H., Lee, K., Lee, Y.-T., Yoo, J., Jeong, G.: IEEE International Solid- State Circuits Conference (ISSCC) Digest of Technical Papers 2012, pp. 46–48 (2012)
31. Chen, S., Gibbons, P., Nath, S.: Proceedings of CIDR, 2011, p. 21 (2011)
32. Kang, M.J., Park, T.J., Kwon, Y.W., Ahn, D.H., Kang, Y.S., Jeong, H., Ahn, S.J., Song, Y.J., Kim, B.C., Nam, S.W., Kang, H.K., Jeong, G.T., Chung, C.H.: IEEE International Electron Devices Meeting 2011, pp. 3.1.1–3.1.4 (2011)
33. Lung, H.L., BrightSky, M., Chien, W.C., Wu, J.Y., Kim, S., Kim, W., Cheng, H.Y., Zhu, Y., Wang, T.Y., Cheek, R., Bruce, R., Lam, C.: IEEE VLSI Technology Digest, pp.118–119 (2014)
34. Kaneko, Y., Tanaka, H., Ueda, M., Kato, Y., Fujii, E.: IEEE Trans. Electron Dev. **58**, 1311 (2011)
35. Muller, J., et al.: Ferroelectric hafnium oxide: a CMOS-compatible and highly scalable approach to future ferroelectric memories. In: IEEE IEDM Technical Digest, pp. 10.8.1–10.8.4 (2013)
36. Shiga, H., et al.: A 1.6GB/s DDR2 128 Mb chain FeRAM with scalable octal bitline and sensing schemes. In: IEEE International Solid-State Circuits Conference Digest of Technical Papers, pp. 464–465, 465a, February 2009

37. Cypress: 4-Mbit (256 K × 16) F-RAM Memory. Cypress, San Jose (2016)
38. Cypress: 64-Kbit (8 K × 8) Serial (SPI) Automotive F-RAM. Cypress, San Jose (2015)
39. Taito, Y., Nakano, M., Okimoto, H., Okada, D., Ito, T., Kono, T., Noguchi, K., Hidaka, H., Yamauchi, T.: A 28 nm embedded SG-MONOS flash macro for automotive achieving 200 MHz read operation and 2.0 MB/s write throughput at Tj, of 170 °C. In: ISSCC 2015, pp. 132–133 (2015)
40. Wong, H.-S.P., Ahn, C., Cao, J., Chen, H.-Y., Fong, S.W., Jiang, Z., Neumann, C., Qin, S., Sohn, J., Wu, Y., Yu, S., Zheng, X.: Stanford memory trends. https://nano.stanford.edu/stanford-memory-trends. Accessed 5 Jan 2016
41. International Technology Roadmap for Semiconductors: Semiconductor industry association (2013)

Chapter 4
Normally-Off Computing

Takashi Nakada and Hiroshi Nakamura

Abstract Normally-off computing is based on synergetic effect of aggressive power gating and non-volatile memory. To maximize opportunity of power reduction, as frequent as possible power management is necessary but too frequent state transition increases overhead energy. To manage this tradeoff, Break Even Time is introduced. If BET is small, there is more opportunity of power reduction. Therefore, new device that has small BET is desired. In addition to this, architectural technologies are also important to reduce effective BET by combining multiple devices that has different characteristics. Finally, using such heterogeneous devices, dynamic behavior is managed by software and energy consumption is minimized. To realize normally-off computing, hardware/software co-design and co-optimization are required. This is a key for not only system designers, but also hardware engineers and software developers.

Keywords Power management · Break even time · Design methodology

4.1 Introduction

In this chapter, we introduce design methodology of normally-off computing to build ultimate low-power system with several technologies that are explained in previous chapters.

The most important problem to be solved is to minimize energy consumption while satisfying certain performance constraint. In general, any low-power techniques have overheads, which may decrease system performance. To realize optimal system it

T. Nakada (✉) · H. Nakamura
The University of Tokyo, Bunkyo, Tokyo, Japan
e-mail: nakada@hal.ipc.i.u-tokyo.ac.jp

H. Nakamura
e-mail: nakamura@hal.ipc.i.u-tokyo.ac.jp

T. Nakada and H. Nakamura (eds.), *Normally-Off Computing*,
DOI 10.1007/978-4-431-56505-5_4

is important to minimize overhead and maximize energy reduction with appropriate combination of suitable techniques.

When designing low-power systems and their optimal management, the most important points are Break Even Time (BET), which explained in Chap. 2. BET is an important index which indicates when and which low-power techniques should be used.

BET is determined by the amounts of energy overhead and energy reduction. Therefore, from a viewpoint of device technologies, new devices whose BET is smaller than that of existing devices should be desired. As new generation non-volatile memories realize smaller BET than that of traditional non-volatile memory, they are good candidate for normally of computing systems.

Additionally, architectural technologies also help to reduce effective BET. By combining different types of devices and architectural management can realize smaller BET than that of monolithic devices. For instance, hierarchical, heterogeneous and hybrid structure are good candidate.

Meanwhile, modern computer systems are getting more complicated. They consist of a wide variety of modules, which have different characteristics. As a result, generic performance and energy models are strongly required to develop an efficient power management scheme for these systems.

Though the respective power managements are different depending on the target hardware and software, the basics of the design methodology are common. We will summarize the basics with practical examples in following part.

To realize such optimal power management, cooperation and cooptimization of different design layers, including algorithm, OS, compiler, architecture, circuit and device, are definitely required.

4.2 Device Technologies

For energy efficient computing, low power devices are fundamental. In this section, we take a general view of device technology. To realize normally-off computing, aggressive low power technologies such as power gating and new generation non-volatile memories are indispensable.

As we introduced in Chap. 3, many kind of new generation non-volatile memories (NVRAMs) are evolved. When compared with volatile memories such as SRAM, NVRAM is almost comparable. But still requires larger access energy especially for write access and longer access latency. From a viewpoint of BET, if the difference of access energy between volatile memory and NVRAM become smaller, the BET also becomes smaller. As a result, in many situations, NVRAMs are superior to volatile memories. Therefore, non-volatile memory that has smaller BET is desired for normally-off computing.

As we mentioned in Chap. 2, the fine-grain power management are well studied and realized. To obtain the maximum energy reduction, it is important to make full usage of the fine-grained power management.

Additionally, other low power device technologies that can shorten BET are also important. For effective power gating, fast and low energy overhead power switches are desired. Such overheads depend on circuit capacity. This circuit capacity is proportion to circuit area. Thus, fine-grained power gating can reduce overheads and shorten BET. As a result, opportunity of energy reduction is maximized.

4.3 Architectural Technologies

From a viewpoint of device technology, perfect device does not exist. Because of many restrictions, such as physical phenomena, material characteristics, manufacture cost and so on, almost all devices have tradeoff between energy consumption and performance.

With these different types of devices and architectural technologies, optimized systems can be constructed. A simple example is a cache system. Modern computer systems have hierarchical cache systems and realize effectively fast and huge memory space. In the same way, effectively energy efficient and high performance system can be realized.

In the rest of this section, we explain how to realize optimized memory and hardware systems by architectural technologies.

4.3.1 Memory Hierarchy

For energy efficient computing, hierarchical memory is helpful. However, traditional memory hierarchy considers tradeoff between access speed and capacity. When we consider energy efficiency new memory hierarchy is necessary.

To improve energy efficiency, sophisticated combinations of the non-volatile and the volatile memories would be a better solution. Namely, it is worth considering to utilize the fast and small volatile memories for performance sensitive parts and the non-volatile memories for the rest. This solution includes integration of volatile and non-volatile memories within a layer (Fig. 4.1b) unlike the traditional configuration which integrating with uniform memory in each memory layer (Fig. 4.1a).

Moreover, high attention should be paid to data movement between memory layers. In conventional memory hierarchy, hardware manages data movement between cache memory and main memory, whereas system software manages between the main memory and the external memories. In contrast, in the new memory hierarchy (Fig. 4.1b), memories with different speed are used within the same layer. Additionally, non-volatility makes a paradigm shift in the cache system. Traditional cache systems assume their contents cannot keep without power supply. However, non-volatile cache memory can maintain their contents without power supply. This advantage may cause drastically changes in the cache systems.

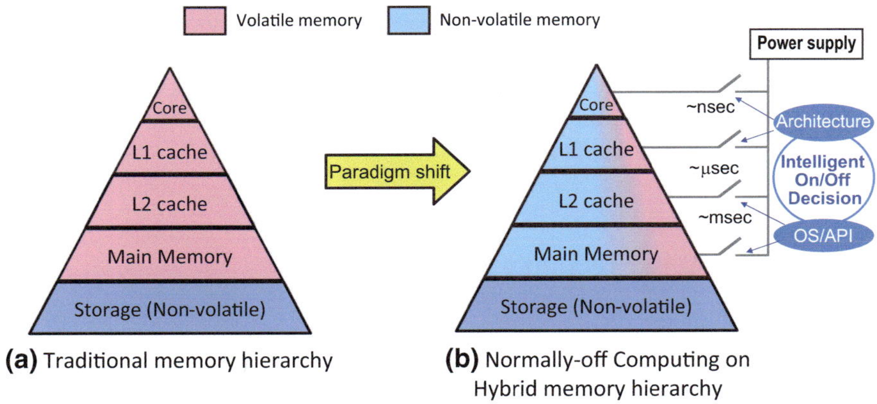

Fig. 4.1 Hybrid memory hierarchy

Thus, more sophisticated memory management is required. For example, it would be worth considering that even in the same memory layer the volatile memory is managed by hardware and the non-volatile memory is managed by system software.

4.3.2 Heterogeneous Hardware

As we briefly introduced in Chap. 2, heterogeneous hardware is useful to adopt a variety of performance requirement. In general, rich or large hardware can achieve high performance, but energy efficiency is low. On the other hand simple or small hardware is energy efficient, but low performance. Additionally, power on overhead is important. For richer or larger hardware requires larger power on overhead due to its larger charging capacity and more complicated initialization process. Therefore, there is also tradeoff between these different types of hardware. For instance, rich hardware is suitable for high power requirement, and simple hardware is good for low performance requirement.

To minimize total energy consumption, an adaptive hardware selection is important. Simple tradeoff is already explained in Chap. 2. However, we only considered an independent task. When there are multiple tasks, optimal scheduling solution become complicated.

As a result, device and architecture technologies provide way to realize normally-off computing. To establish normally-off computing, software supports that manage activities are necessary.

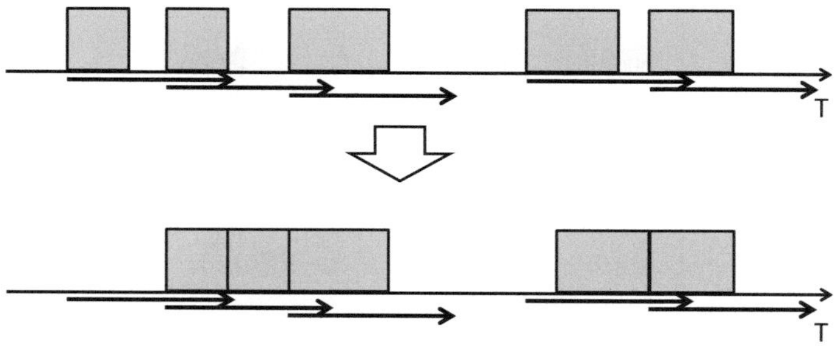

Fig. 4.2 Temporal activity control

4.4 Activity Control Technologies

To minimize energy consumption, activity and granularity controls are important. In general, there is flexibility of scheduling, as long as performance requirements are maintained. Therefore, any activities that are not performance critical should be rescheduled to maximize effectiveness of low power technologies.

From viewpoint of spatial granularity, it is effective to minimize active modules and completely turn unused modules off. From viewpoint of temporal granularity, once an operation is started, as much operations as possible should be done continuously to maximize the length of idle time.

If the length of idle time becomes longer successfully, lower energy consumption power mode can be chosen and the total energy consumption will decrease.

Such cooptimization can be implemented both software and hardware. As we mentioned in Sect. 4.2, hardware implementation can minimize management overhead, but its behavior is given by predefined logic. Conversely, software implementation can manage more flexible, but control overhead is relatively larger.

By hardware, data management such as caching and buffering are good solution. Hardware can manage with very fine granularity and low overhead. However their adaptability may be limited because their functionality is fixed at time of manufacturing. If several kinds of hardware are implemented heterogeneously, different management mechanisms or characteristics can be selected at runtime.

By software, optimization via task and data scheduling is possible. For task scheduling, execution timing and execution order are important. Software can realize more flexible management than hardware but management overhead is relatively high.

Figure 4.2 shows an example of temporal activity control. Gray boxes show tasks and their width shows execution time. Black arrow shows arrival time (start point) and deadline (end point) of each task. If the length of the arrow is longer than the width of the box, such task can be reschedulable. In this example first three tasks and next two tasks can be executed continuously respectively. As a result, several short idle times are merged and become one longer idle time. If the length exceeds BET, deeper sleep mode can be chosen and the total energy consumption decreases.

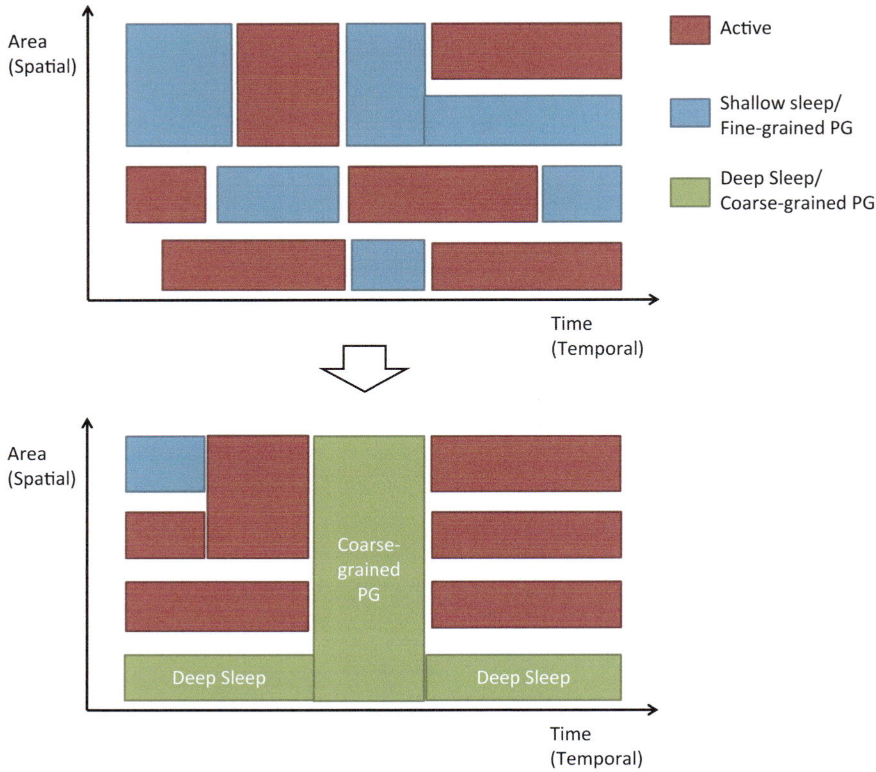

Fig. 4.3 Spatiotemporal granularity management

Furthermore, granularity management can be extended to spatiotemporal space as shown in Fig. 4.3. In this figure, x-axis shows time domain and y-axis shows spatial domain. To simplify the illustration, spatial domain is divided into only four. Red boxes indicate active domains and the rest is idle period. Before optimization, activities are scattered and idle periods are small and short. Thus only fine-grain PG and shallow sleep can be chosen (blue boxes). On the other hand, after granularity optimization, activities are gathered and idle periods are large and long. As a result, coarse-grain PG can be applied and Deep Sleep can be chosen (green boxes).

Additionally, heterogeneous hardware can be well adapted to dynamic behavior such as execution time variation and thus contribute to improve the energy efficiency. High performance but energy hunger hardware is necessary for heavy workload to maintain performance. Low performance but energy efficient hardware is important to reduce total energy consumption.

4.5 Normally-Off Computing Design Methodology

To design normally-off computing systems, it is important to understand each component and their power management technologies. There are several important feature values such as granularity, activity, break even time and so on. When these values are changed, the optimal power management may be changed. Therefore, it is important to figure out what kind of optimization changes these values. Then interaction between neighbor components can be expressed these feature values. As a result, system wide optimization is possible.

Firstly, BET should be as small as possible. Latest power management continuously shortens BET by technology improvement. It is indispensable to understand BETs of each hardware component. Based on these BETs, following power management is considered.

To minimize total energy consumption, energy breakdown should be observed. Then components that consumes large amount of energy should be optimized. Basic analysis is definitely important to understand their behavior and characteristics especially granularity or activity.

If idle time which is longer than BET is found, suitable power management such as PG should be applied. As we mentioned, contents in volatile memory will be lost when PG is applied. To cope with this problem, there are two options; one is data migration, the other is to replace with non-volatile memory.

In many cases, replacing with non-volatile memory is good candidate. Thanks to its non-volatility, reexecution or store/restore cost will be zero, though its access energy is slightly larger.

Additionally, neighbor component may affect their activities. Therefore, their interaction should be considered. For example, non-volatile buffer is good candidate to control their activities. Meanwhile, trade-offs between volatile memory and non-volatile memory or overheads of additional activity management modules should be managed.

Using this methodology, we will explain detailed power management technology of each component in Chap. 5. Specifically, logic, cache memory, micro-controllers, task scheduling and data management will be introduced. Then, some practical systems that introduced normally-off computing will be introduced in Chap. 6. Specifically, healthcare system, mobile information device and Sensor network system for smart city application will be introduced.

Chapter 5
Technologies for Realizing Normally-Off Computing

Takashi Nakada, Shinobu Fujita, Masanori Hayashikoshi, Yoshikazu Fujimori and Hiroshi Nakamura

Abstract Normally-off computing relies on device technologies, architectural and activity management technologies. To realize normally-off computing systems, details of each technology should be studied. Device technologies provide component that has as small BET as possible by minimizing energy overhead. Architectural technologies combined several devices that have different characteristics and realize flexible components that can adapt to a wide variety of applications. Activity management is done by software such as task scheduling. Based on these knowledges, normally-off computing system is realized.

Keywords Low power technology · Logic · Cache · Micro controller · Scheduling

T. Nakada (✉)
The University of Tokyo, Bunkyo, Tokyo, Japan
e-mail: nakada@hal.ipc.i.u-tokyo.ac.jp

S. Fujita
Toshiba Corporation, Minato, Tokyo, Japan
e-mail: shinobu.fujita@toshiba.co.jp

M. Hayashikoshi
Renesas Electronics Corporation, Koto, Tokyo, Japan
e-mail: masanori.hayashikoshi.cj@renesas.com

Y. Fujimori
ROHM Co., Ltd., Kyoto, Kyoto, Japan
e-mail: yoshikazu.fujimori@mnf.rohm.co.jp

H. Nakamura
The University of Tokyo, Bunkyo, Tokyo, Japan
e-mail: nakamura@hal.ipc.i.u-tokyo.ac.jp

T. Nakada and H. Nakamura (eds.), *Normally-Off Computing*,
DOI 10.1007/978-4-431-56505-5_5

5.1 Overview

In this chapter, detailed technologies for normally-off computing in each component are introduced. In Chap. 4, we introduced device, architectural and activity control technologies.

As we mentioned before, to realize normally-off computing utilize non-volatile memory, temporal and spatial activity management is important.

Before discussing whole system optimization, we should understand details of each component or technology. In each of them, physical phenomenon is different. Then management policies are different.

Particularly, we introduce low power technologies for logic as a device technology. Then cache memory and micro-controllers are explained as architectural technologies. Finally, task scheduling and data management are introduced as activity control technologies. These technologies have different temporal and spatial granularities. For example, logic is basic component and can be controlled with the finest spatial granularity. Therefore power management of the logic can be done by hardware. On the other hand, task scheduling is implemented by software and control with a coarse granularity.

For low power logic circuits, it is important to minimize energy consumption of flip-flops (FFs). To realize FF-level nonvolatility, we introduce hybrid FFs from volatile FF and non-volatile memory. Since FFs are broadly located in logic circuit, bit by bit organization is the most suitable. Thus, a non-volatile FF consists of a traditional volatile FF and one bit of non-volatile memory. This non-volatile FF can be replaced with any volatile FF whatever we want and this organization is scalable.

For low power caches, it is important to minimize energy especially write energy. To solve this issue, hybrid configuration is very useful. Total energy consumption of the hybrid cache is strongly depends on their management policy.

For low power processors, heterogeneous multiprocessors are good candidate. Energy efficiency of processors are depends on their performance. Thus, lower performance processors can achieve better energy efficiency and higher performance is required to meet performance requirement. Additionally, modern processors have multiple sleep modes. Deeper sleep mode can reduce their energy consumption during sleep state. On the other hand, shallower sleep mode realizes quick restart with small energy overhead. To minimize total energy consumption under performance restriction, adaptive processor selection and adaptive data management are important.

Storage and network systems are out of scope, because these systems need specific optimization technologies.

5.2 Logic

5.2.1 Non-volatile Logic

In this section, we introduce a Non-volatile Logic (NVL) which integrated with traditional flip flop (FF) and ferroelectric (FE) capacitor. Firstly, characteristics of FeRAM, which is ferroelectric (FE) capacitor based memory, are introduced. Secondly, the NVL is introduced. Finally, normally-off computing with NVL is explained.

5.2.1.1 Characteristics of FeRAM

When an access frequency for a memory macro is low, the use of an FeRAM is suitable to reduce power consumption of a normally-off system because standby current of the FeRAM can be cut off by turning off the power supply.

Figure 5.1 shows relationship between an access frequency and the power consumption of each memory macro under the same process (130 nm LSTB), where the x-axis and the y-axis indicate the number of the word written into the memory macro per second and the average of the power consumption, respectively. In case of Normally-Off, power supply is turned off during interval of writes every single second. When the access frequency is low, the power consumption is dominated by the standby power caused by off-leakage current of a transistor and power loss during the power turning on and off. If it becomes high, the power consumption is determined by the dynamic power of memory access (proportional to quantity of data). When the number of the word becomes lower than 1k words, the power consumption of the FeRAM becomes lower than those of an SRAM and a FLASH memory.

5.2.1.2 Design and Implementation of Non-volatile Logic

A non-volatile logic (NVL) technology is very suitable to implement a logic circuit on a normally-off computing system because the non-volatile logic circuit has capability to retain a logic state without power supply. In the NVL, all registers consist of non-

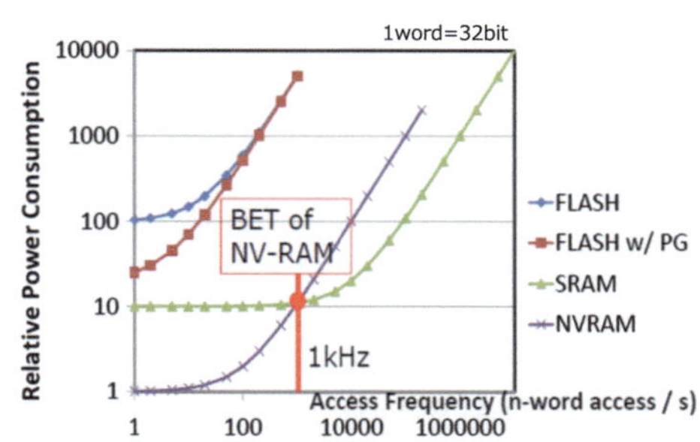

Fig. 5.1 Relationship between an access frequency and power consumption

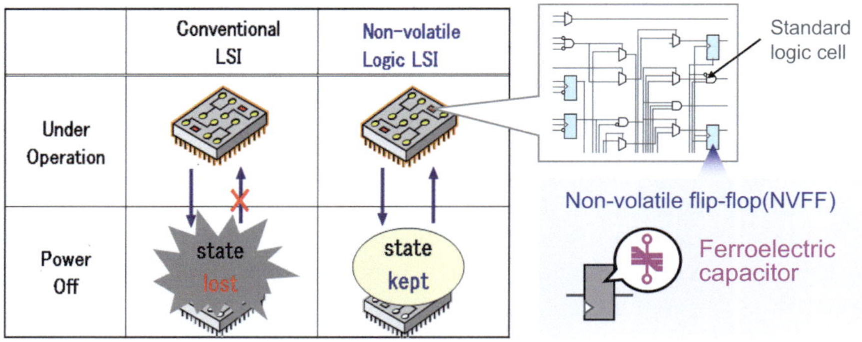

Fig. 5.2 Non-volatile logic (NVL) technology

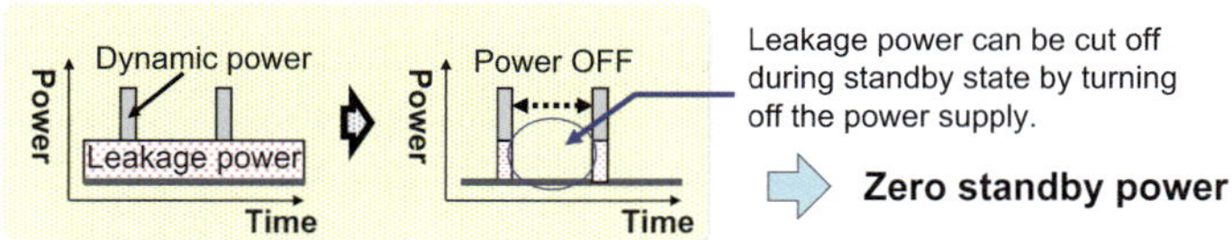

Fig. 5.3 Zero standby power by turning off power supply of an NVL circuit

volatile flip-flops (NVFFs) which have non-volatile storage capability by using a ferroelectric (FE) capacitor as shown in Fig. 5.2. The FE capacitor is a kind of a non-volatile storage device, which is used in a memory cell of an FeRAM as described in Sect. 3.2. Therefore, it is possible to cut off the power supply during standby state, and standby power can be reduced to zero as shown in Fig. 5.3. An arbitrary logic circuit is easily nonvolatized by replacing all conventional flip-flops (FFs) to NVFFs.

An important point of NVL circuit design is to implement an NVFF without performance degradation in comparison with a conventional FF. If an FE capacitor implemented in the NVFF acts as an additional load device in the logic operation, it causes large loss of the dynamic power. Moreover, read and write operations of the FE capacitor should be performed by CMOS-logic-compatible voltage supply to prevent area overhead caused by an additional power line for operations of the FE capacitor.

To overcome above issues, the NVFF is designed on the basis of a conventional FF circuit with FE capacitors as shown in Fig. 5.4. During a logic operation, FE capacitors in the NVFF are isolated from the conventional FF circuit by disabled access driver as shown in Fig. 5.5. Therefore, performance of the NVFF is almost as same as that of the ordinary FF during a logic operation.

Since the FE capacitors are accessed only when power supply turns on or shut off, access times of FE capacitors are same to the cycles of power ON/OFF cycle. Therefore, regarding the fatigue endurance of 10^{12} cycles of FE capacitors, the NVL is able to bear a 10 year long performance assuming power ON/OFF happening 3000 times in a second. The access sequence for FE capacitors in all NVFFs is controlled by an NVFF controller, which is used to detect turning on/off of the power supply.

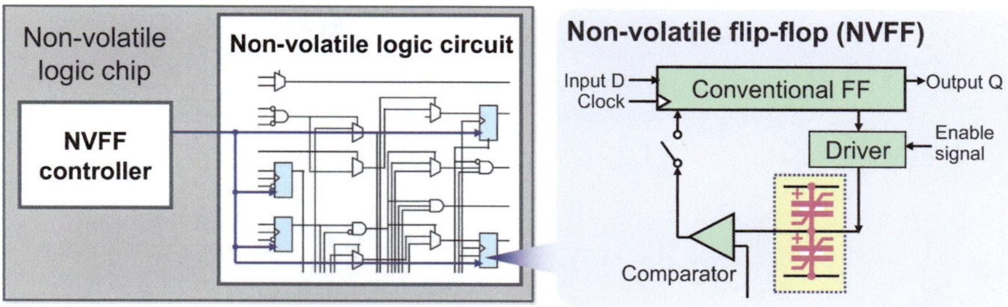

Fig. 5.4 Overall structure of a non-volatile logic circuit

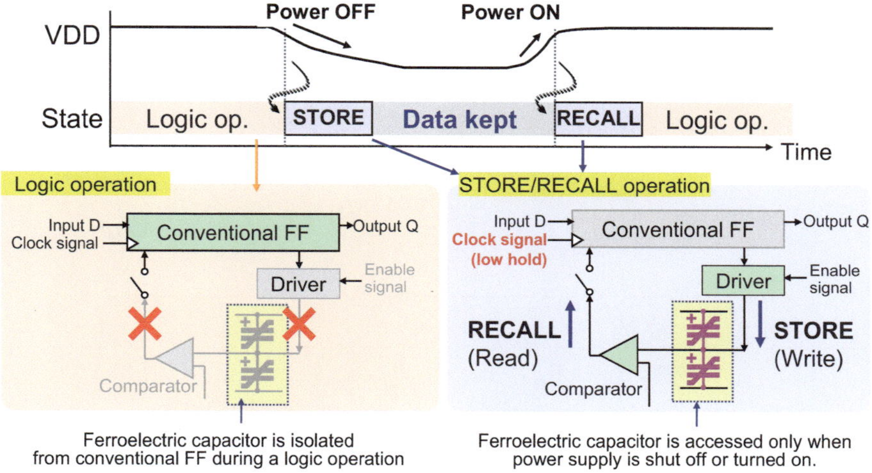

Fig. 5.5 Basic operation of a non-volatile logic circuit

Moreover, the use of complementarily stored data in coupled FE capacitors makes it possible to implement a compact NVFF circuit with wide read voltage margin. Its storage capability can guarantee 10 years retention at 85 °C without power supply after CMOS-logic-compatible voltage supply (VDD = 1.5 V).

In the conventional read circuit of the FE capacitor, a remnant polarization charge corresponding to stored data S is read out by capacitive coupling effect between the FE capacitor and a load capacitor. Since capacitance of the FE capacitor depends on S as shown in Fig. 5.6, a read voltage signal V_{OUT} also does. This circuitry is widely used in an FeRAM because a load capacitor can be easily implemented by parasitic capacitance on a bit line. However, it requires higher voltage than CMOS logic circuit to achieve a read voltage margin ΔV_{OUT} for 10-years data retention capability. Moreover, the load capacitor causes large area overhead in the NVFF.

On the other hand, the use of coupled FE capacitors with complementary data storage makes it possible to achieve a read voltage margin ΔV_{OUT} for 10-years data retention capability with CMOS-logic-compatible voltage operation

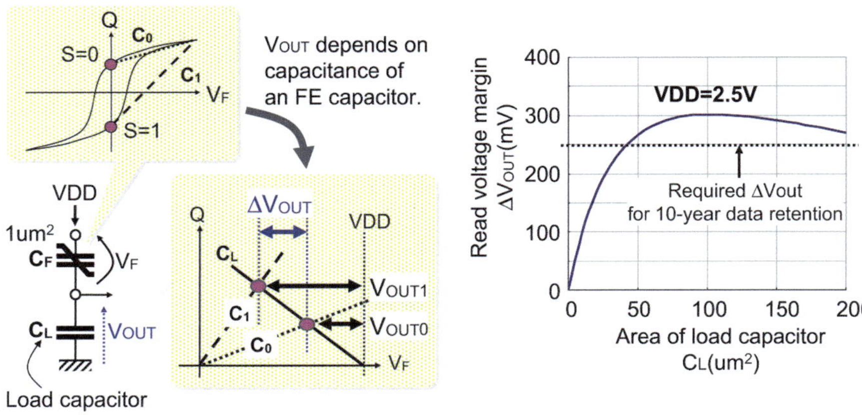

Fig. 5.6 Conventional read circuit of the ferroelectric capacitor

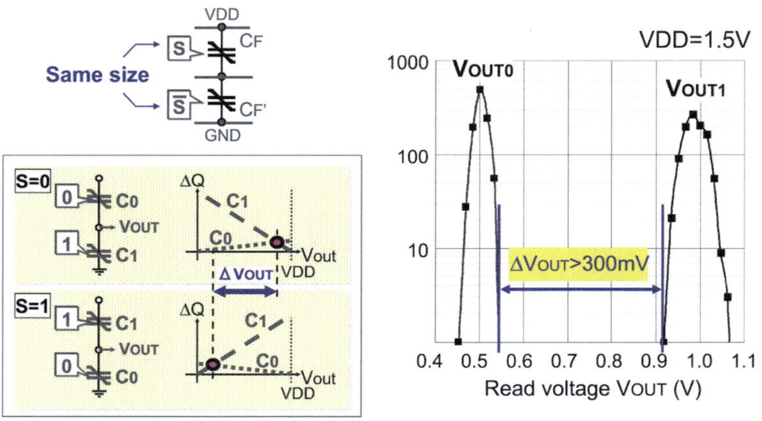

Fig. 5.7 Ferroelectric capacitive coupling with complementary data storage

(VDD = 1.5 V), because change of capacitance ratio between two series-connected capacitors becomes larger than that of a conventional read circuit as shown in Fig. 5.7. Moreover, this technique contributes to implement the compact NVFF without large area overhead of load capacitor.

5.2.1.3 Normally-Off Computing for Non-volatile Logic

In the normally-off systems, the use of NVL technology leads to superior power reduction than non-volatile memory-based implementation because data transfer between the FF and the ferroelectric capacitor is localized in the NVL circuit, which contributes to reduce power dissipation caused by data transfer between the FF and the ferroelectric capacitor as shown in Fig. 5.8.

When the power supply turns off, stored data in the FF is transferred from the FF to the ferroelectric capacitor so as to retain stored data during power off state. As a same, written data in the ferroelectric capacitor is recovered to the FF after power supply turns on. Hence, in the normally-off system, it is important to reduce power dissipation caused by data transfer between the FF and the ferroelectric capacitor as shown in Fig. 5.9.

Figure 5.10 shows the comparison between FeRAM-based implementation and NVL-based one, where one-hundred store and recall cycles are performed per second. In the FeRAM-based implementation, stored data in FFs has to be transferred to the ferroelectric capacitor in an FeRAM cell through long data path. Moreover, basically

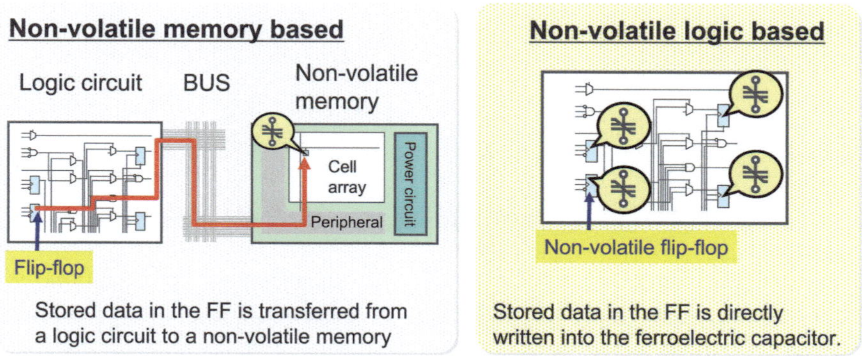

Fig. 5.8 Non-volatile memory-based implementation and NVL-based implementation

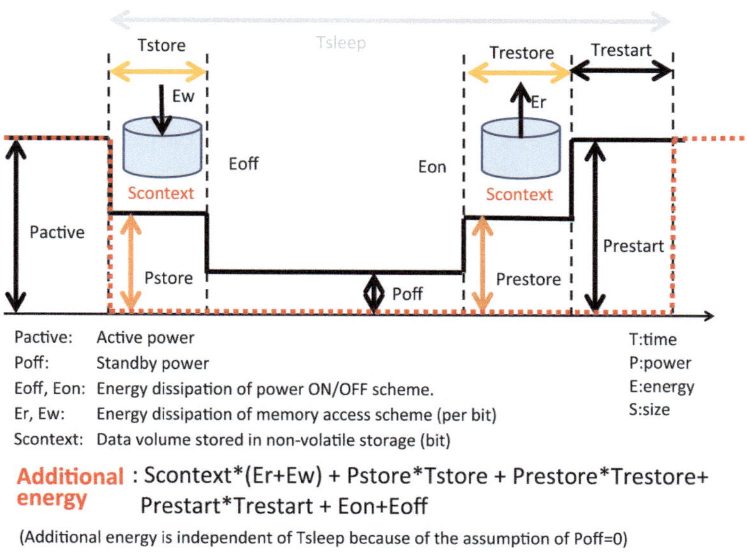

Fig. 5.9 Estimated power dissipation in the power control scheme

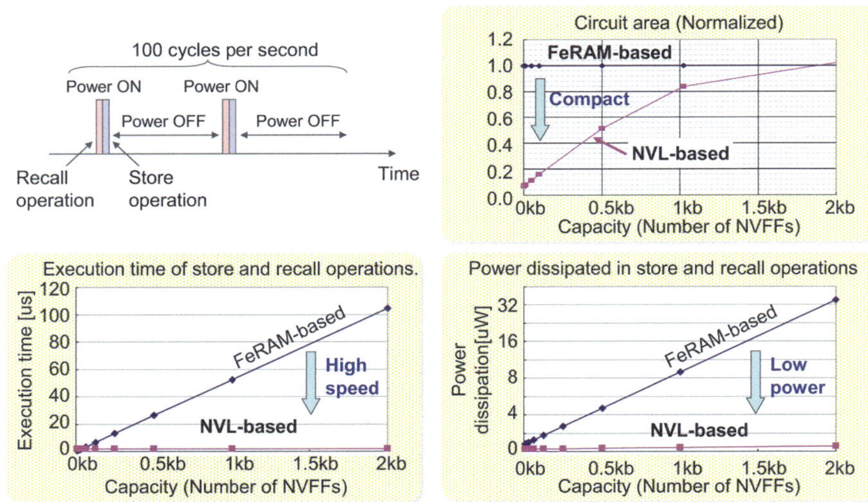

Fig. 5.10 Comparison of FeRAM-based implementation and NVL-based one

FeRAM needs higher voltage than CMOS logic circuit in the access scheme, which causes large loss of the dynamic power.

On the other hand, in the NVL circuit, data transfer between the FF and the ferroelectric capacitor can be performed without long data path because it is localized in the NVFF circuit. Moreover, the operation voltage of the NVFF is compatible to CMOS logic circuit. Therefore, NVL-based circuit can reduce large loss of dynamic power caused by data transfer between the FF and the non-volatile storage element. Basically, dynamic power dissipated in a store and recall cycle of the NVL-based circuit is always lower than that of the FeRAM-based one independent of volume of stored data.

The localized data transfer between the FF and the non-volatile storage element in the NVFF also contributes to reduce an execution time of the store and recall cycle because all NVFFs perform data transfer scheme simultaneously during the store and recall cycle. That is, a period of data transfer scheme is always constant independent of the number of NVFFs.

Figure 5.10 shows that when a number of FFs used in the NVL circuit is lower than 2 k cells, area of the NVL circuit becomes smaller than that of FeRAM-based implementation because the NVL circuit needs no additional data bus or the power control unit to generate high voltage supply.

The design flow of the proposed NVL circuit is same to that of the conventional one except for non-volatile replacement as shown in Fig. 5.11. A given Resister-Transfer Level (RTL) specification is synthesized into a gate netlist together with the NVFF controller module. Then, all conventional FFs in a target logic module are replaced by NVFFs. At the same time, signal wires used to transit NVFF control signals are connected to the NVFF controller and NVFFs. Since the performance of

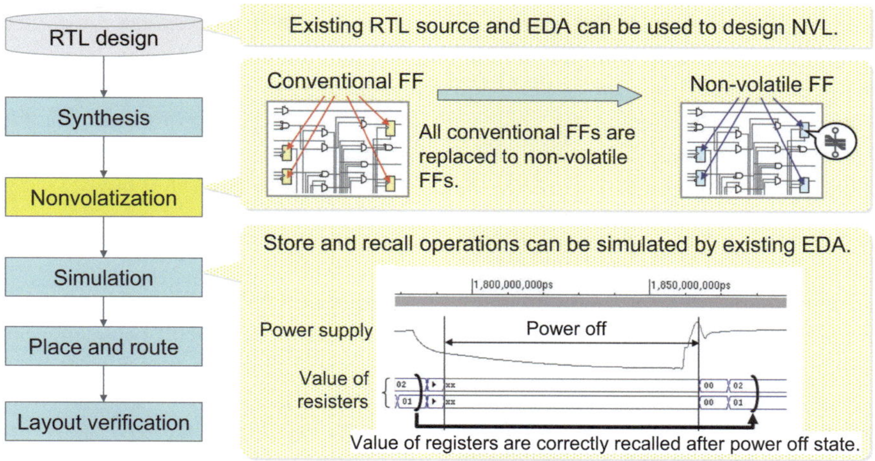

Fig. 5.11 Design flow of an NVL circuit

the NVFF is the same as that of the conventional one in a logic operation, timing mismatch will not occur after the NV replacement.

As a design example of the NVL circuit, an NVMCU using 1.3k NVFFs is described in Sect. 6.2, which is contribute to reduce total power dissipated in a normally-off ECG-SoC application.

5.3 Memory/Cache

5.3.1 Embedded STT-MRAM for High-Performance Processors

5.3.1.1 Memory Core Design for Normally-Off Operation

Advanced microprocessors use the control of power supply to each circuit block, i.e., the power gating (PG) technique, to decrease the static CMOS power. However, power gating cannot decrease static power of volatile memories such as flip-flops or SRAM, as data in these memories are lost if PG is applied. To decrease the static power further, nonvolatile memories have been expected to be used instead of volatile memories. The most promising candidate among nonvolatile memories recently developed is the spin-torque transfer magnetic random access memory (STT-MRAM) using magnetic tunnel junction having perpendicular magnetic anisotropy (p-MTJ) [13, 14]. This has high speed reaching that of SRAM and practically unlimited memory endurance, as described in Chap. 3. However, it should be noted that if the write energy or other active energy of MRAM is larger than total energy of the standby leakage power, it is difficult to reduce total energy consumed by MRAM compared with that of SRAM, and as a result, there is no benefit in employing MRAM instead of SRAM. Therefore, write programming energy (PE) is a key factor for reducing the leakage power of SRAM and CPU or SoC. The PE is the product of the write current, voltage and time for write operation. The energy reduction by replacing SRAM with MRAM is roughly expressed as follows:

$$\text{leakage current} \times \text{Vdd} \times \text{standby time (SRAM)} - \text{write current} \times \text{Vdd} \times \text{write time (MRAM)}. \quad (5.1)$$

To enable the energy reduction, the write operation of MTJ with both fast speed and low power simultaneously is needed. However, there is a tradeoff between write current and write time; no MTJs with both fast speed and low power have been reported. Based on recent reports [14, 29], perpendicular MTJ achieved write time of 3 ns with low current of 30 μA, resulting in small PE of less than 0.09 pJ.

This MTJ is generally used in one transistor (1T)-1MTJ memory cell. The read speed of 1T-1MTJ cell is not as fast as that of SRAM, since resistance change ratio of MTJ is as small as about 2–3 times. To solve this problem, nonvolatile cross-coupled inverters serially connected with two p-MTJs were proposed [15–17]. These cross-coupled inverters are called "nonvolatile latch". A nonvolatile latch with two p-MTJs can be used as a 6T-NV-SRAM cell as shown in Fig. 5.12. It is advantageous that the number of transistors is the same as that of a conventional 6T-SRAM cell. Based on a precise physical layout, it has a 28% area overhead.

In a normal mode, the 6T-NV-SRAM cell writes data (WRITE) and reads data (READ) like a conventional SRAM cell. Data in the 6T-NV-SRAM cell are stored (STORE) in two p-MTJs before power shutdown. When the power is turned on again, the cell data are automatically recalled (RECALL) from two p-MTJs. Each operating condition is shown in Fig. 5.13. The two p-MTJs are set to the AP-state to initialize the 6T-NV-SRAM cell before WRITE (RESET). For this RESET operation, two MTJs in the 6T-NV-SRAM cell are programmed to high-resistance (antiparallel (AP)) state by applying low-level signal (L) to BL and BLB. The conditions in WRITE and READ are the same as those for conventional SRAM. The simulated delay overheads of WRITE and READ due to insertion of p-MTJs are 0 and 27 ps, respectively, in a 45 nm CMOS technology. However, read delay overhead can be neglected in total delay of SRAM.

However, these NV-SRAMs include leakage paths like SRAM, and herein are called "normally-on" memory cells. Although other kinds of NV-SRAM combined with SRAM cells and two MTJs were also proposed [18, 19], as shown in Fig. 5.14, all of them have normally-on memory cells. For normally-on memory cells, rapid PG is needed after every write operation to reduce leakage power. In general processors, cache access frequencies are a few ns for L1, a few tens of ns for L2 and more than 50 ns for L3. Even for L3 cache, highly frequent (>20 MHz) PG is needed to reduce leakage power. Although such highly frequent power switching is theoretically possible, such highly frequent PG increases circuit area greatly and degrades computing performance. Furthermore, reliable and stable power management circuits are hard to implement in real SoCs. Also, considering the cache access scheme,

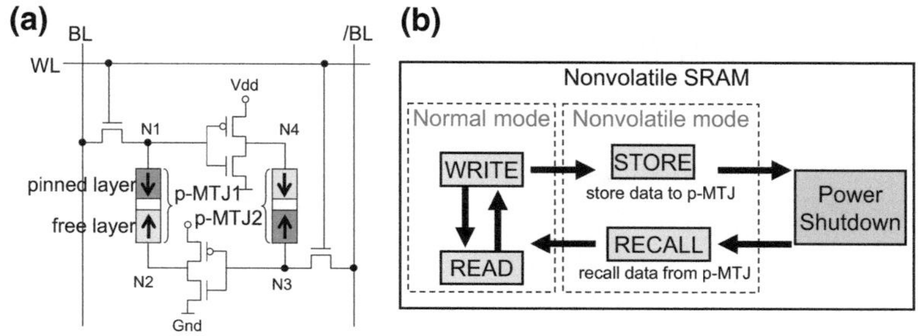

Fig. 5.12 **a** Circuit diagram and **b** operation of 6T-NV-SRAM cell

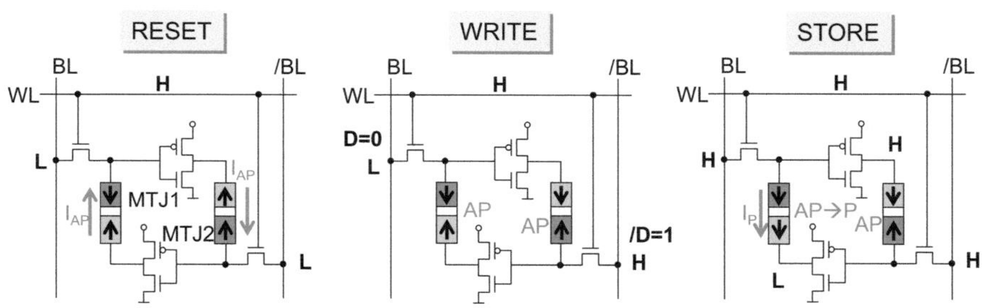

Fig. 5.13 RESET, WRITE and STORE operation conditions

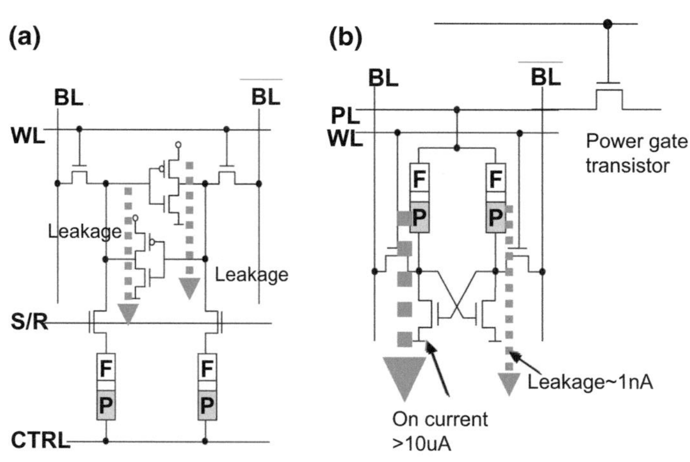

Fig. 5.14 NV-SRAM with **a** 8T-2MTJ [18] and **b** 4T-2MTJ [19]. "F" denotes a free layer and "P" denotes a pinned layer

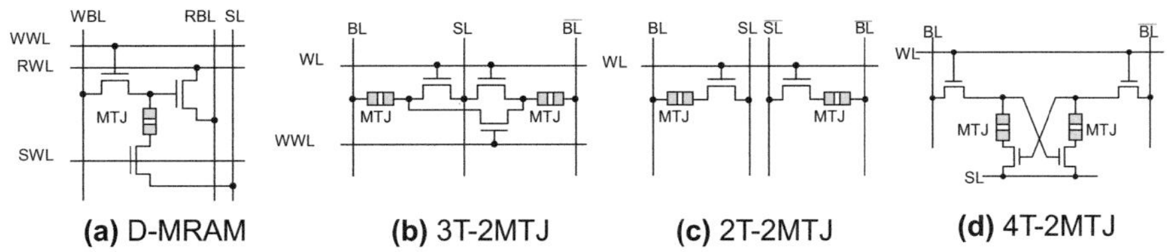

Fig. 5.15 Four kinds of normally-off type memory cell designs using 3T-1MTJ [20], 3T-2MTJ [21], 2T-2MTJ [22] and 4T-2MTJ [23]

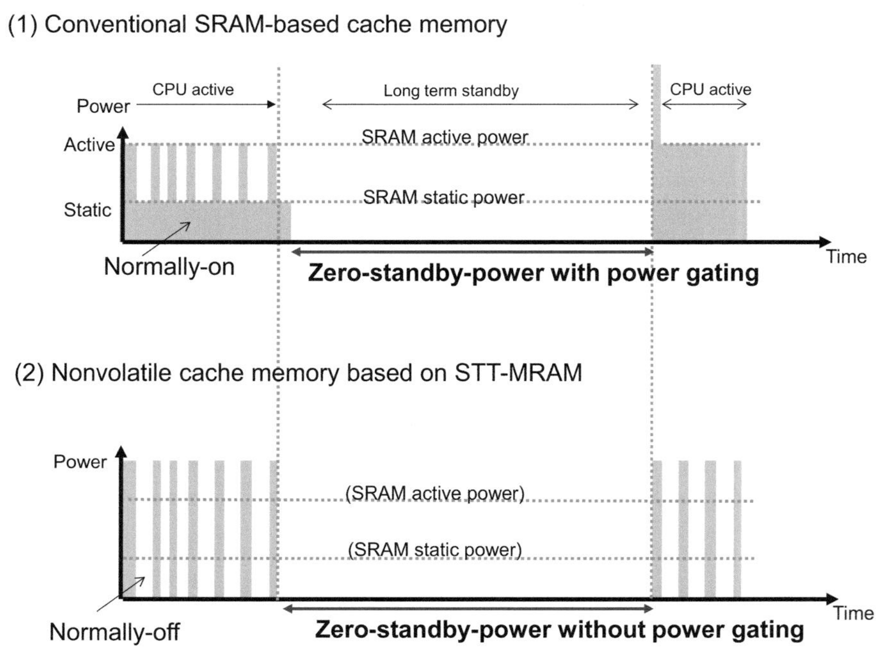

Fig. 5.16 Decrease in leakage power for the short waiting state during active state of CPU by normally-off type STT-MRAM

the power domain should be 64 to 512 bit memory cells for cache memories. In such fine-grain power domain, the memory design concept has to be changed from "normally-on" with PG to "normally-off" without PG that enables effective power reduction even for just one clock-cycle standby state. Figure 5.15 show the novel normally-off memory cells without leakage paths designed [20–23] based on the most advanced p-STT-MRAM. Without leakage paths, leakage power can decrease even for a very short standby state during active state of CPU, as shown in Fig. 5.16. The details of each normally-off memory cell design are described in the next section.

- 2T-2MTJ cell

A fundamental issue of STT-MRAM is that the on-off resistance ratio is as small as ×2 to ×3. Hence, read time with 1T-1MTJ-based memory cell is relatively long (<~10 ns) compared with that of cache for SRAM. In order to achieve faster read operation, the 2-cell/1-bit structure (2T-2MTJ cell) was proposed [22]. Figure 5.15c shows a memory cell topology. In this cell structure, 1-bit data are stored by using two cell units. Complementary data are stored in two 1T-1MTJ cell units. Each cell unit in the memory cell array includes an MTJ device and a selection transistor connected in series. A gate terminal of the selection transistor is connected to wordline WL. One end of the cell unit is connected to bitline BL0 and the other end is connected to source line SL0. One end of the other cell unit is connected to bitline /BL0 and the other end is connected to source line /SL0. Both bitlines and source lines extend in the column direction and one end thereof is connected to write circuit and read circuit via column selector. ON/OFF of column selector is controlled by column selection signal CSL from column decoder circuit.

Unlike in the 1T-1MTJ cell structure, since the sense amplifier circuits amplify the difference in cell current between a pair of MTJs, the reference cells are not needed. The cell size of the dual cell estimated in 65 nm CMOS technology is 0.45 μm^2, which is much smaller than that of SRAM. Read latency can be greatly reduced and the sensing margin is doubled because reference cells are eliminated. In the prior current-based read circuits, the read cell currents flow from Vdd to Vss continuously during read operation. On the other hand, in the proposed sensing scheme, since while the accessed wordline, WL, is activated, the differential bitlines, BL and /BL, are left floating and work as "capacitance", the read cell currents are stored in bitlines, BL and /BL, and no currents flow from Vdd to Vss directly. The sensing margin can be expanded by this "current-integral" scheme that accumulates the small difference between complementary cell currents through "L" resistance MTJ and "H" resistance MTJ. The area for the proposed sense amplifier circuit is 27% less than that for the conventional current sensing circuit. To write the complementary resistance states in a pair of MTJ devices in a 1-bit cell, symmetrical write currents continuously flow according to the write pulse width of the p-MTJ device, from a source line to a bitline and from a bitline to a source line.

- 3T-2MTJ cell

Figure 5.15b shows the proposed schematics and the layout of the cell for normally-off cache memory. It consists of 3 transistors and 2 MTJs. The cell area is 0.504 μm^2, which is almost identical to the 65 nm SRAM cell area. Write is executed by applying the current between blt and blc via the write transistor nw. When the current flows from blt to blc, MTJ at true side and MTJ at complement side become AP (Antiparallel: High-R) and P (Parallel: Low-R) respectively and vice versa. The simultaneous write instead of the sequential time-sharing write [21] is executed in this way. This has become possible using the low-voltage MTJ and the extra write.

The simultaneous write improves the cycle time and write power consumption. Target write current for 125 MHz operation is estimated to be 50 μA using the write

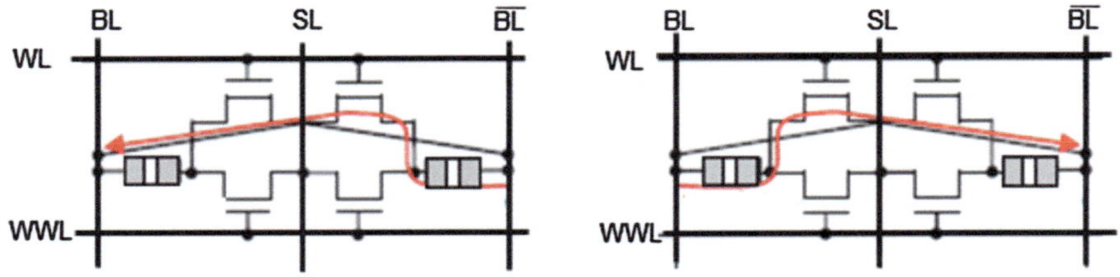

Fig. 5.17 Memory cell schematics suitable for lower voltage write operation

current dependency on the write time. Note that smaller write current is possible by increasing the write pulse width, but it results in the increase of write power consumption. The nominal supply voltage for write operation is estimated to be 1.3 V. If the lower voltage operation is required, it is possible using the memory cell shown in Fig. 5.17. The target maximum read current, which is the upper limit for avoiding the data corruption, is 10 μA.

- 4T-2MTJ cell

It is known that the smallest SRAM cell is a 4-transistor-based design with no load resistance PMOSs, and therefore there are no leakage paths from PMOS to NMOS. This cell design has not been used for real applications, mainly because the data retention is short and the refresh is needed as in DRAM. The authors newly propose 4T-SRAM with 2MTJs that requires no refresh thanks to nonvolatile MTJ memory cell. Figure 5.15 shows the proposed NV-SRAM cell design without the current leakage paths. MTJs are serially connected to signal path between bit line/bit line bar (BL/BLB) and source line (SL) through transistor (Tr.) and metal layer 1 and 3 (M1/M3) and M2/M4.

In the write operation shown in Fig. 5.18, the data that will be written into a targeted NV-SRAM cell are first read out and then written into MTJs. Here, the settings of signals for SL, BL and BLB are Vm, Vmh and Vml (Vmh>Vm>Vml), respectively, if N1="0" and N2="1" were read out. When WL is selected, MTJ1 and MTJ2 are programmed to high resistance in the antiparallel state (AP) and low resistance in the parallel state (P), respectively, by the write current between BL/BLB and SL. After WL is set to a high level, N1 and N2 are set to "0" and "1", respectively. Although the data retention of SRAM itself after the WRITE operation is just about 1 ms, it is not an issue since the data are stored in MTJs.

After the WRITE operation, the voltage level of each 4T-SRAM node is automatically fixed by the combination of two MTJ resistances, as shown in Fig. 5.19, meaning that the memory data automatically shift from MTJs to 4T-SRAM cell. Here, BL and BLB voltage are set to Vpc after SL voltage is 0 V, and then WL voltage swings from 0 V to Vdd during the read operation. If MTJ1 is AP state and MTJ2 is P state, N1="0" and N2="1" are read out because of Rap>Rp. The time when NV-SRAM becomes ready to read after the write operation is as short as 1.46 ns by circuit simulation, which is a great advantage for the high-performance cache

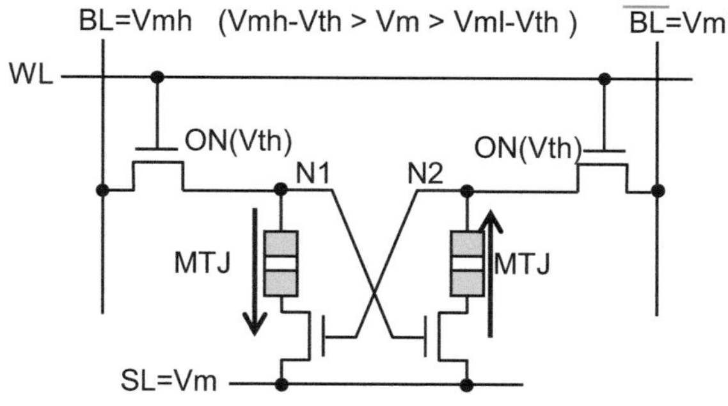

Fig. 5.18 Operation flow for MTJ-WRITE

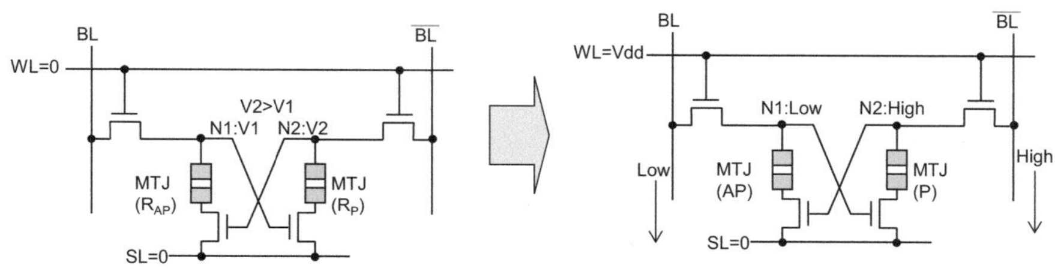

Fig. 5.19 Memory data flow from MTJ to SRAM for READ operation

memory operation. Designed layout of 4T-NV-SRAM suggests smaller cell area than that of 6T-SRAM due to the smaller number of transistors.

- 3T-1MTJ cell (D-MRAM)

DRAM/MRAM hybrid "D-MRAM" with 3T-1MTJ memory cell design (Fig. 5.15) was fabricated. It has high speed and no current leakage path. The capacitance of RTD's gate and RTM's source-drain capacitance (off-state) is equivalent to a capacitor for DRAM. For the DRAM-mode, write and read operation is explained in Fig. 5.20, respectively. The retention time calculated is about 10 us. For the MRAM-mode, write operation is based on current injection into the MTJ through two transistors by the spin torque transfer, as shown in Fig. 5.20, and read operation uses a current sensing scheme, where read signal lines and reference signal lines are switched to connect the sense amplifier according to the read mode (DRAM or MRAM), as shown in Fig. 5.20.

Subarray configuration of 1 Mb D-MRAM was designed for cache memory. Using this memory macro, performance and power were simulated based on low-power 65 nm CMOS technology. As for the MTJs, an advanced perpendicular (p-) MTJ STT-MRAM having fast speed and low power (write time: 3 ns, write current: 30 μA) was used [14]. The programming energy, 0.09 pJ, is the smallest ever reported. The write access time of DRAM- and MRAM-mode is 1.5 and 4.5 ns, and these read access times are both about 2.2 ns, indicating that DRAM-mode has higher memory access speed than MRAM-mode. DRAM-mode also has lower memory access energy than MRAM-mode.

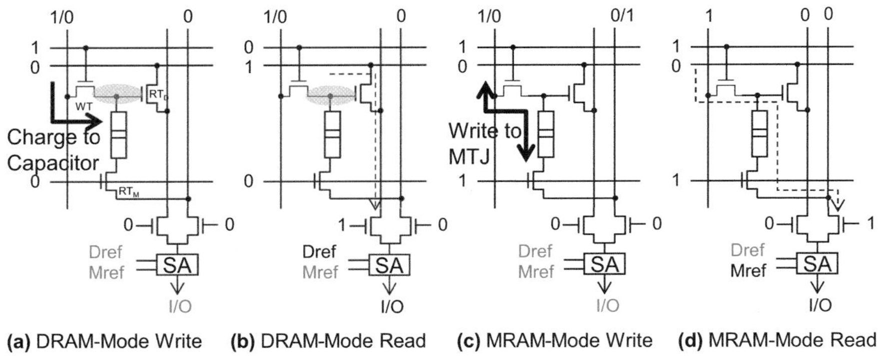

Fig. 5.20 Operation of D-MRAM. **a** The DRAM-Mode Write operation drives WWL high to store 1/0 in the cell capacitor. **b** The DRAM-Mode Read operation drives RWL high, and then RBL current rises or remains because charging up through RTD depends on cell capacitance. The difference compared to reference for DRAM-Mode (Dref) is read in sense amp (SA). **c** The MRAM-Mode Write operation drives WWL and SWL high to switch MTJ state (AP/P). **d** The DRAM-Mode Read operation drives WWL and SWL below voltage of MRAM-Mode Write operation. The difference of SL signal caused by the difference of MTJ resistance state is read in SA with reference for MRAM-Mode (Mref)

Thus, when a D-MRAM-cache line is initially accessed, it works as DRAM and the data are retained with refresh operation every retention time, ~10 us. However, since refresh energy has to be consumed in the DRAM-mode memory, it is suitable that the data are transferred to the MRAM-mode after the time to change from DRAM-mode to MRAM-mode, Tc. To determine the Tc, it should also be considered that MRAM-mode has larger write power than DRAM-mode. Tc also depends on cache access pattern determined by the application running on the CPU. In our case study considering cache access analysis, 50 us was selected for Tc. The memory-mode (DRAM-mode/MRAM mode) is selected for each cache line, as both write and read access speed simulated is as fast as 2 ns. Then, when the second access to the same cache line by cache hit, the cache system rewrites the same data in the cache line by the DRAM-mode, and data remain for another retention time. When there are several (N-times) cache hits, the data are written into MTJs by the MRAM-mode on the cache line in the background, and the "flag" data on the line is changed from "0" to "1", and then D-MRAM cells start to operate as MRAM after this situation. The number N was defined by:

$$\text{DRAM rewrite energy/bit} \times 2N > \text{MRAM write energy/bit}. \tag{5.2}$$

On the other hand, when the cache lines are overwritten by other higher-priority data, the flag data is changed back to "0". This scheme is very effective to reduce energy for cache memory.

5.3.1.2 Memory Macro Design for Normally-Off Operation

Although leakage power has been eliminated at the memory cell array area, there is still large leakage power in the peripheral circuits for high-speed STT-MRAM. It has to be decreased for realizing completely normally-off NVRAM. For this purpose, high-speed power gating (HS-PG) for the peripherals is required, as standby state of high-speed STT-RAM is very short ($<\sim$50 ns). Fine-grain power domain for the STT-MRAM has been adopted [24], as shown in Fig. 5.21; power domains are divided into four blocks of 256 kb local memory subarrays, decoders, read drivers, write drivers, and sense amplifiers. Further, all power supply lines are classified into local and global power domains. Power is supplied to the local memory subarray power domain, only when the subarray is accessed, which can be judged from the access address. On the other hand, the global power domain does not have such "address dependency" of PG. Circuit area and active current differ from circuit to circuit, and power supply lines were adjusted to sufficient strength and durability depending on their active current and circuit area. To minimize area and delay overhead due to HS-PG, the relation between granularity and delay overhead was analyzed. According to this analysis, 32 KB power domain is the best to achieve sub-4 ns access time within area overhead acceptable for the cache application. In this proposed power gating scheme, during active states, fine-grained power gating of the local power domain effectively reduces static energy consumption, and for long standby states the power gating of the global power domain can reduce total standby energy more effectively. To detect the long-standby states, we adopt timer-based state transition. In Fig. 5.22, the state 0 represents read or write operation state, state 1 represents non-operation state where every local power domain is powered off, and state 2 represents global power gated state in which state every local power domain and global power domain is powered off. The on-off repetition is dramatically different between local power domain ($<$ 1 ns) and global power domain ($\sim$20 ns). In state 1, the power-gated

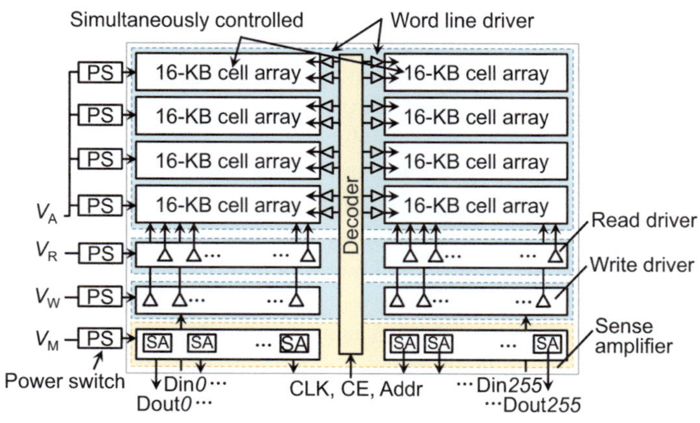

Fig. 5.21 Block diagrams for power domains

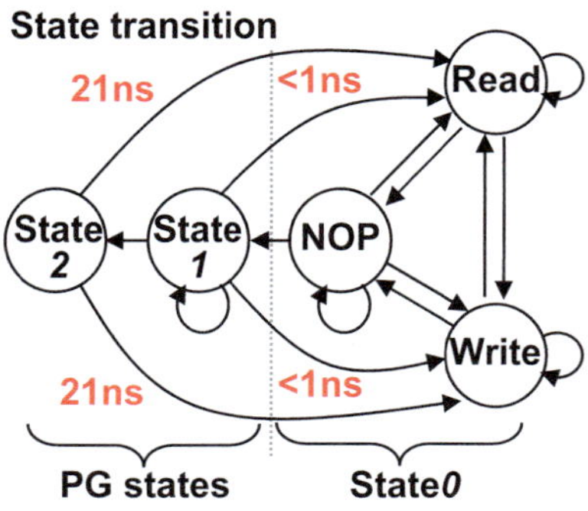

Fig. 5.22 Five states for STT-MRAM memory macros and the state transition to lower power states

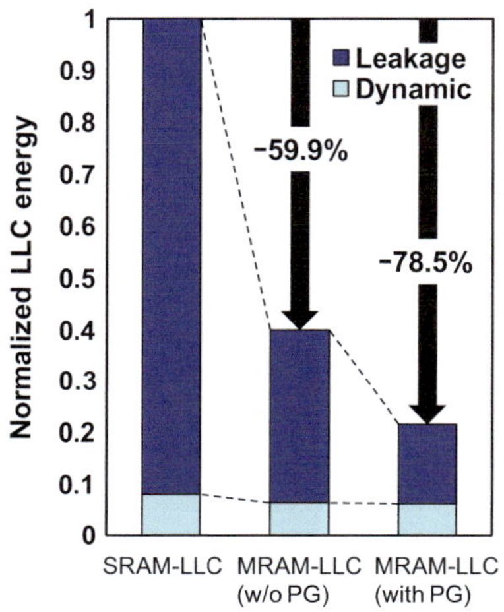

Fig. 5.23 Comparison of LLC energy between SRAM-based LLC and MRAM-based ones with and without power gating(PG)

domain can be quickly powered on. In state 2, the power-on process needs several cycles and the on-off energy overhead is negligible, although state 2 can reduce standby power more effectively than state 1.

Power and performance in processor with 2T-2MTJ STT-MRAM-based cache memory for LLC was evaluated based on CPU simulator customized using Gem5 [25]. The CPU core uses 3-way issue out-of-order ARMv7-based single-core architecture. Processor simulations were conducted with SPEC CPU2006 benchmarks [26]. Our proposed STT-MRAM, compared to the typical 0.8 V operating SRAM design, can reduce the energy per instruction (EPI) of the total cache memory by 78.5%, as shown in Fig. 5.23, while keeping the instruction per cycle (IPC) performance degradation within 6% in spite of its latency overhead. On the other hand, the EPI for other STT-MRAMs previously reported is greatly increased. For our proposed STT-MRAM, I/O width can be expanded to 256 bit or 512 bit needed for the cache-line width (e.g. 64 B) thanks to low-power operation, whereas previous ones are less than 64 bit as listed in Table II. It should be noted that decrease in power for CPU active state is attributable to decrease in leakage power of L2 cache memory.

5.4 Micro Controller

5.4.1 Overview of sensor node

As an example of equipment which utilizes micro controllers, we introduce sensor nodes. Firstly, their architecture is explained. Secondary, their energy model is considered.

1. *Architecture of current sensor node*

The conventional sensor node is consists of sensor modules, and microcontroller (Fig. 5.24).

Power is always supplied to the sensor modules and microcontroller. And the processing of sampling, average and etc., of sensor data is performed periodically at microcontroller according to the request from systems, and the processing data is sent out to sensor networks through the sensor-net interface.

2. *Possibility of power reduction for the sensor node*

In the smart community, the sensors used in environment system (temperature, brightness, motion sensors and etc.) are performed at the data sampling intervals of 50 ms∼10 s. Since power is always supplied to sensor modules and microcontroller (MCU), the unnecessary power is consumed during the period that has not been processed (Figs. 5.25 and 5.26).

The challenges to realize the low power consumption of above sensor nodes,

- To supply power at sensor modules and microcontroller on when needed,
- To reduce the active currents of microcontroller.

Therefore, it is necessary to minimize the overhead power and time at power return, and reduce the processing load of the microcontroller and maximize the power-off time as long as possible, and also minimize the power-on area (Fig. 5.27).

In order to realize low power sensor node, we propose a normally-off architecture of microcontroller, aiming to achieve a normally-off computing system in future sensor nodes.

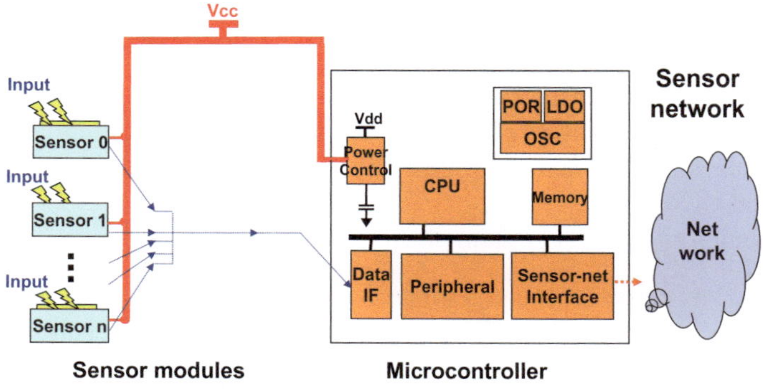

Fig. 5.24 Architecture for the conventional sensor node

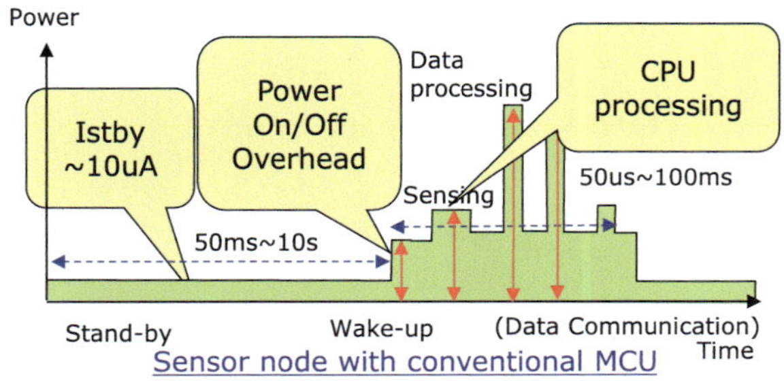

Fig. 5.25 Transition of processing and power consumption in sensor system by microcontroller

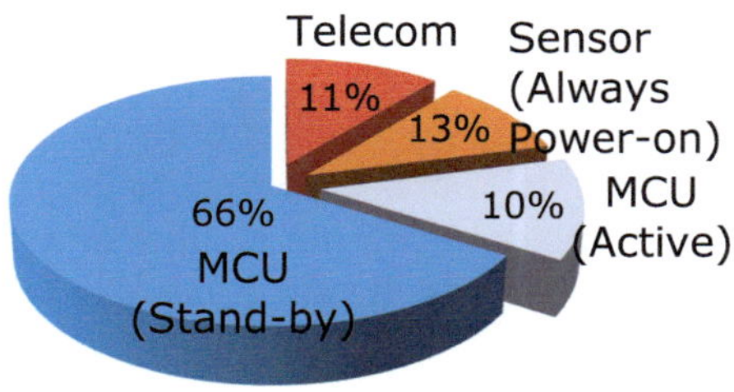

Fig. 5.26 Distribution of power consumption of the sensor system

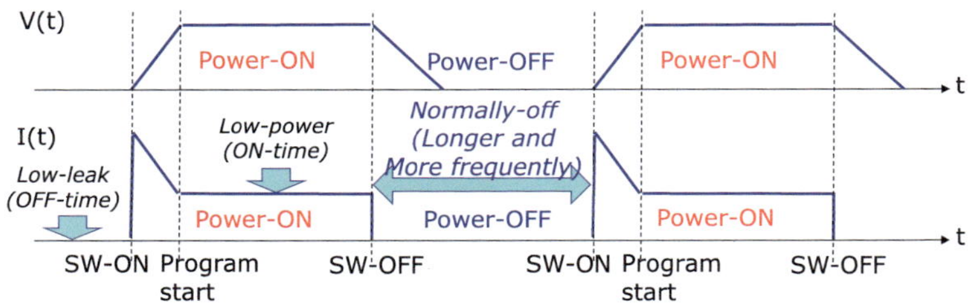

Fig. 5.27 Waveform of intermittent operation

5.4.2 Low-power technologies

(1) Autonomous standby mode transition technology

Now, the development of low-power technology of microcontroller (MCU), various attempts have been made, and are classified as below (Fig. 5.28).

- Reduction of the CPU activating time by the low-power / high-performance CPU.
- Minimization of standby power by the standby mode setting of multiple.
- Minimization of CPU activating time and CPU start-up cycles by the improved peripheral circuit.
- Minimization of wasted operating time by the CPU start-up time reduction.

These are low power technologies for efficient use of standby mode. The purpose is to reduce the power consumption energy, by allowing transition period as long as possible in lower power standby mode and by reducing of the overhead time of the state transition of standby and active.

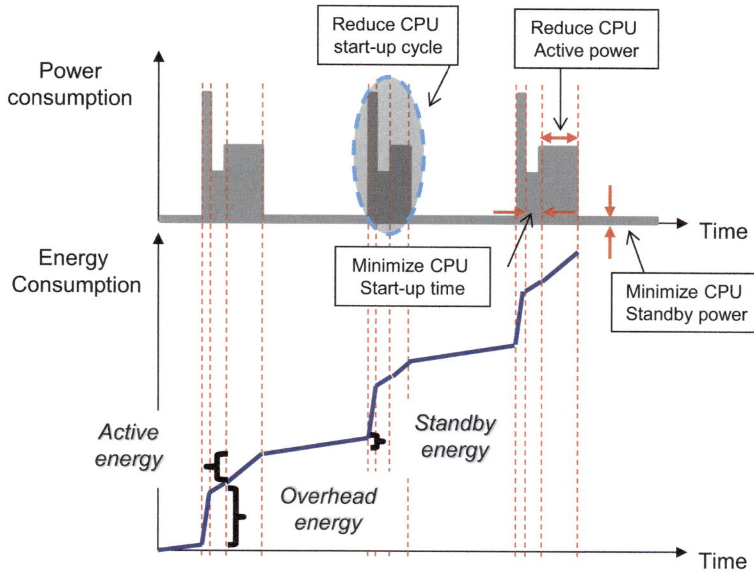

Fig. 5.28 Power consumption transition of CPU

The software development is performed while checking the power consumption of each operation in real time by using the development tools in order to minimize the power consumption energy. However, it is difficult to select the optimum standby mode in consideration of the dynamic changes in condition and environment of use.

In order to maximize the effect of low-power consumption technology of MCU, the autonomous standby mode transition technology is to make select and transition the optimum standby mode autonomously to minimize power consumption energy by quantifying the constraints of hardware and software.

Understanding of the constraint condition of hardware side and software side is needed in order to select the optimal standby mode. The constraint of the hardware side is the breakeven time (BET), which is derived from standby power consumption specific to each device and the energy overhead of the state transition of the standby and active. The constraint of the software side is the waiting time depending on the application.

The power consumption energy in the standby state and the energy overhead of returning from the standby state are schematically described about supported standby mode in microcontroller (Fig. 5.29). Some standby modes are existed in microcontroller. One case is that energy consumption in standby mode is very low as power off, but the overhead energy at return from standby mode is very large. Other case is that the return from standby mode is fast, but power consumption energy in standby mode is not reduced as IDLE mode. Each standby mode has different BET. When the waiting time at the request of the software is more than BET, it is better to make a transition to a deeper standby mode. In this work, the optimal standby mode

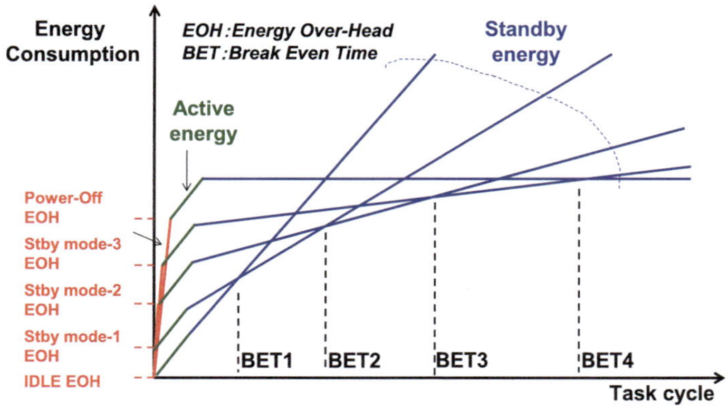

Fig. 5.29 Break-even time of each standby mode

can be selected autonomously by normally-off power manager, which include the autonomous standby mode selection scheme.

(2) Heterogeneous Multi-processors

Another important technique to minimize energy consumption is heterogeneous multi-processors. As we mentioned in previous section, there are BETs for multiple stand-by modes. In addition to this technology, heterogeneous configuration can support wider range of applications.

The power consumption of heterogeneous multi-processors is illustrated in Fig. 5.30. In this figure green boxes show normal mode and hatched triangles show energy overhead of state transition between normal and sleep states.

This figure clearly shows that their overhead energy of state transition is in proportion to their performance. Namely, smaller CPU has low performance but overhead energy for power cycle is small, vice versa. As a result, the small CPU is good for low load execution and the large CPU is good for high load execution. Therefore, adaptive CPU selection is important.

However, to realize this adaptive execution, it is important to know each task size. In general, hardware does not know the task size and CPU selection is done by a task scheduler.

One of the most important roles of hardware is to minimize overhead and to provide wide range of CPU performance. Then, the task scheduler can assign tasks to suitable CPU.

5.5 Scheduling

In the previous sections, we introduce low power hardware techniques. In addition to them, it is important to software techniques to minimize total energy consumption.

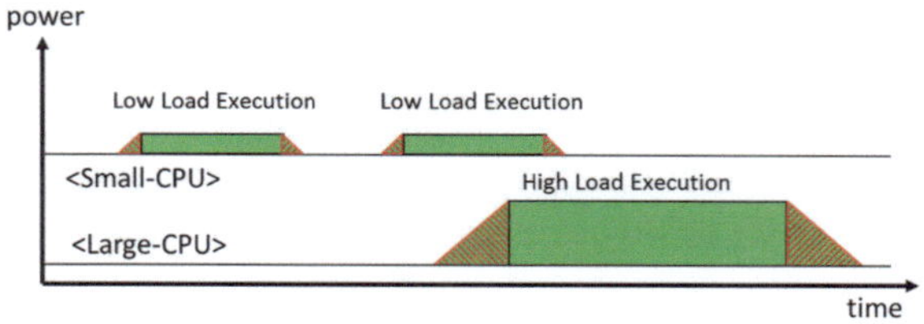

Fig. 5.30 Power consumption of small and large-CPUs

In this section, an energy efficient task scheduling [8] is introduced. Since the energy consumption of the processor is larger than that of data access in most applications, data management is optimized after the task scheduling.

5.5.1 Introduction

As technology advances static power increases more rapidly than dynamic power [12], and now it gets comparable to dynamic power consumption. As static power is consumed without any contribution to computing, its reduction is strongly required. A wide variety of power reduction techniques has been proposed and realized, including clock gating, power gating, DVFS, and so on.

To make full usage of these techniques, task scheduling is a key but for multi-core systems it is very complicated due to the huge exploration space. Most previous work assumes that the deadline equals to the execution interval to simplify this problem. Under this assumption, only one execution interval should be considered and use its solution repeatedly. This strategy still satisfies the application requirements. On the other hand, this strategy limits the effectiveness of the low power techniques, due to a smaller design space. For further power reduction, it is necessary to make full usage of the original deadline. For example, sensing or video/audio streaming applications allow longer deadlines by buffering, though the throughput requirements are strict.

In this section, we realize a method to find an optimal design that achieves the best energy efficiency under constraints such as the deadline. Our ultimate goal is to find the most energy efficient design for a periodically executed embedded system under deadline and other constraints with a reasonable cost. The design includes not only a hardware design such as core selection but also a software design such as scheduling and power management. Especially when the deadline is longer than the execution interval, we search a larger design space to make full use of the long deadline.

5.5.2 Scheduling for Multi-core System

In the case of several kinds of processors executing multiple tasks, there exist many combinations of assignments.

In general more powerful processor core consumes more energy. There exists an empirical model between them called Pollack's Rule [11]. According to this model, the performance is roughly proportional to the square root of a processor's area. The static power is proportional to the area while the dynamic power is more complex and it can be regarded as being roughly proportional to the performance since switching rates differ between functional units and other parts (in general, they switch less frequently than FUs). Therefore, using cores that are as small as possible is the best from the viewpoint of energy efficiency.

In a real scheduling problem, there are many constraints such as deadlines. To relieve this deadline constraint, we adopt pipeline scheduling. If there is only one core in the system, the total execution time of the task should be shorter than the input interval regardless of the deadline. On the other hand, if we have multiple cores and the task can be divided into subtasks, we can assign each subtask to different cores. Then, the assigned task size of each core is smaller and we can use a smaller and more energy efficient core. Theoretically, a task can be divided into the same number of subtasks as the quotient of the deadline divided by the input interval and these subtasks can then be executed on a multi-core processor having the same number of cores. However, in real systems, we should consider a parallelizing overhead such as the communication delay.

After task division and assignment to cores, we adopt *Lumped execution*. If the deadline is longer than the input interval, we can buffer several input data that are used multiple instance. Then several subtasks are executed continuously. At the same time, idle periods also appeared continuously. In other word, though the activity ratio is not changed, the length of each idle period becomes larger and power mode transition becomes less frequent. As a result, there exists a better opportunity of energy reduction.

In this section, to simplify the discussion, we assume that all the output of each subtask should be stored in an external shared memory. Namely, all communication between cores is done via the memory.

Finally, multi-core scheduling is expressed by the following variables. Obviously, our goal is to find the optimal values, which minimize the energy consumption, for these variables.

- Number of cores
- Core and memory selection
- Task and core mapping
- Scheduling in each core

5.5.2.1 Co-optimization

For global optimization, every variable should be considered at the same time, due to their mutual dependence. For example, an optimal number of cores depends on the type of the selected processor core. The optimal processor cores depend on the task scheduling and DPM. DPM and the scheduling depend on each other.

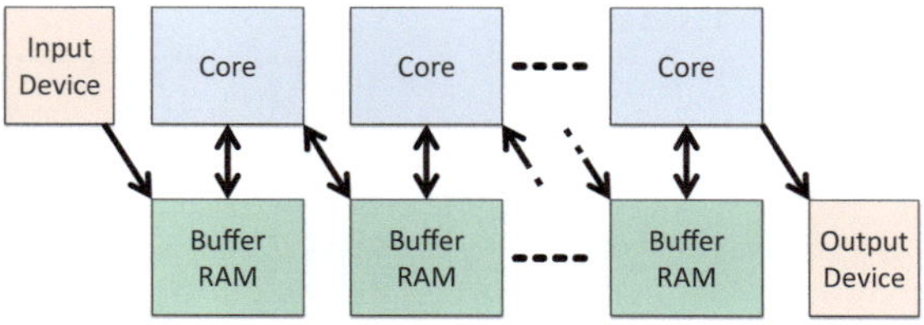

Fig. 5.31 Target system

As a result, especially for multi-core systems, the search space to find an optimal design can easily explode. To solve this problem, an efficient search method is strongly required.

5.5.3 System Description and Constraint Modeling

In this section, we describe our target system in order to find an optimal design. First, we briefly overview our target system, then explain a detailed description for each part.

5.5.3.1 Target System

Our target multi-core embedded system is shown in Fig. 5.31. This system consists of a multi-core processor, buffer memories, input device and output device. The multi-core processor consists of heterogeneous cores that support DPM.

For buffer memories, we assume there exists a tradeoff between the access energy and the access speed, that is, a faster memory consumes a larger energy. As mentioned in Sect. 5.5.1, an input device such as a sensor is integrated with a small memory controller and stores data into the buffer memory by itself. Thus the input data is periodically available in the input buffer memory.

A target application is called a task. We assume the task has already been divided into several subtasks by a programmer and the subtasks have sequential dependency from input to output. The task is invoked repeatedly. We also assume the execution time of each subtask is fixed and already measured for any candidate core.

5.5.3.2 Description Parameters

Description parameters of the target system are categorized into the following parts.

- Task
- Hardware component
- Scheduling with pipelined and lumped execution

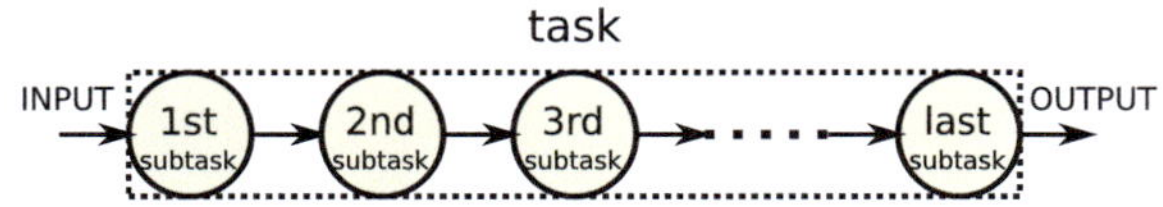

Fig. 5.32 Task and subtasks

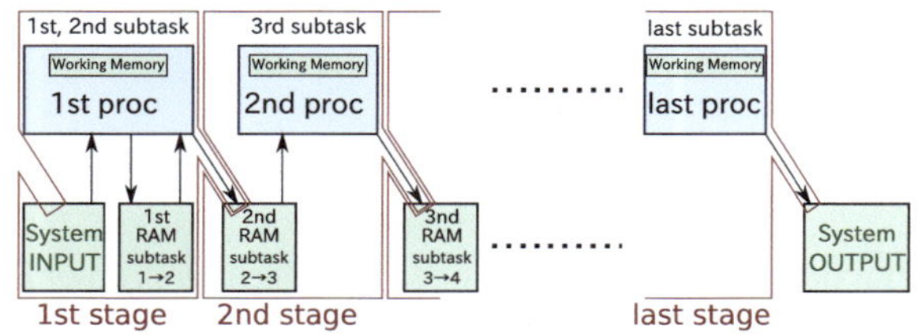

Fig. 5.33 Hardware model

In these parameters, the task and the hardware component parameters are given. On the other hand, to find an optimal solution of the scheduling, which is expressed with some parameters, is our goal. In other words, the task and the hardware parameters are input and the scheduling parameters are output of the multi-core system design.

Each parameter is described in the following sections.

Task Description

As previously mentioned, the target task has been divided into several subtasks. The divided task is shown in Fig. 5.32. Each subtask has an ID s. The first subtask has $s = 1$, and the last subtask has $s = s_{max}$.

All parameters related to the task are as follows. $\{s_n\}$ is a set of subtask IDs in task. T_{In} and T_{Dl} are input interval and total deadline of task respectively.

Hardware Description

The outline of our target hardware is shown in Fig. 5.33. We divide the total execution into several stages. Each stage is executed by one processor core. The core executes one or more than one successive subtasks. For communication between subtasks, a dedicated buffer area is assigned. We assume each stage has its own buffer memory.

All parameters related to the hardware are as follows.

$P_{pStby}[p]$ and $P_{pStat}[p]$ are static power of processor core p in stand-by mode and in active mode respectively. $E_{pDpm}[p]$ is overhead energy when PG is applied to processor core p. $E_{pDyn}[p][s]$ and $T_{proc}[p][s]$ are dynamic energy and processing time of a subtask s on processor core p respectively.

As we mentioned, all power and execution time related parameters have to be measured in advance.

Scheduling Description

As we mentioned in Sect. 5.5.2, the target task is divided into several tasks and some of them are assigned for one processor core. In this section, we describe a scheduling within a core.

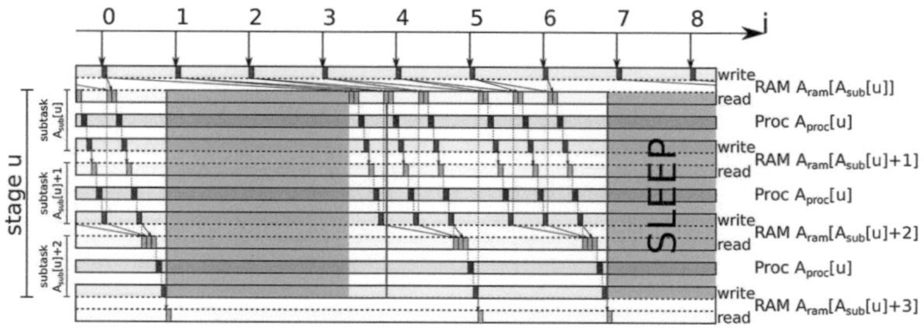

Fig. 5.34 Lumped scheduling

All parameters related to the scheduling are as follows.

- u_{max} # of stages
- $A_{sub}[u]$ The first subtask ID of stage u
- $A_{proc}[u]$ Core ID of stage u
- $A_{ram}[s]$ Memory ID for subtask s's input
- $N_{Lmp}[u]$ # of instances of lumped execution in stage u

Here $A_{sub}[u]$ represents task assignment. Namely, from subtask $A_{sub}[u]$ to $A_{sub}[u+1]-1$ are assigned to stage u. $N_{Lmp}[u]$ represents the number of lumped instances. Figure 5.34 shows an example scheduling when $N_{Lmp}[u] = 2$. In this example, 7 inputs are required for the lumped 2 instances. When the last input (i=6) is available, every subtasks related to the input are executed continuously. Other preceding tasks should be scheduled earlier properly.

Since our design space exploration is done offline and its result is static, when the best configuration of these parameters are found, hardware and software configurations are statically determined. The hardware configuration is directly determined from u_{max}, $A_{proc}[u]$ and $A_{ram}[s]$. The task assignment is determined from $A_{sub}[u]$. Finally, the task scheduling is determined from $N_{Lmp}[u]$. When each core wakes up and starts execution is easily and statically calculated from these scheduling parameters and the task parameters.

5.5.4 *Optimal Scheduling*

5.5.4.1 Design Space

In the previous section, we have listed all input parameters and search variables. The search variables are u_{max}, $A_{sub}[u]$, $A_{proc}[u]$, $N_{Lmp}[u]$.

Finding an optimal design is equal to finding the best solution of these variables.

5.5.4.2 Dynamic Programming

To alleviate the huge computation complexity problem, we propose a sophisticated algorithm using dynamic programming.

First we quantize the execution time t, namely $t = n \cdot \Delta t$ ($n \in \mathbb{N}$). In this equation, Δt is the quantization step. We call each possible assignment of the variables *design* of stage u and assign ID $1, 2, \ldots, c_{max}(u)$. When searching the best design from every possible design for stage $1, \ldots, u-1$ and design $1, \ldots, c$ for stage u and the total latency is shorter than t, take the lowest energy consumption $E_c(t, u, c)$. Also, when using design c for stage u, take the total delay $T(c)$ and the energy consumption $E(c)$. Here, $T(c)$ is also quantized, namely $T(c)$ is rounded up to the nearest $n \cdot \Delta t$. Then, the following equation is established.

$$
\begin{aligned}
E_c(t, 0, 0) &= 0 \\
E_c(t, u, c) &= \min(E_{cAdd}, E_{cOther}) \\
\text{where} \quad & E_{cOther} := \min_{c' < c} E_c(t, u, c') \\
& E_{cAdd} := E(c) + E_c(t - T(c), u - 1, c_{max}(u - 1))
\end{aligned}
$$

Here, E_{cOther} corresponds to the energy when the design c is not used, and E_{cAdd} corresponds to the energy when the design c is used. Based on this equation, the optimal design of the target system is found by dynamic programming.

As we mentioned, t is quantized, the total number of E_c is a polynomial of $T_{Dl}/\Delta t$, u_{max}, c_{max}. This restriction helps to reduce the computation complexity by providing a solution close to the optimal.

As a result, the computation complexity of this algorithm is given by

$$O(S \cdot M \cdot u_{max}/\Delta t).$$

This strategy successfully avoids the exponential computation complexity related to u_{max}. Note that S still has an exponential complexity on s_{max}.

5.6 Data Management

5.6.1 Introduction

For processing cores, Power Gating (PG) [5] can successfully reduce their static energy. On the other hand, applying PG to working memory is difficult because data will be lost if power is not supplied. More specifically, some temporal data in the working memory are still alive even during idle periods. For example, time series analysis and feedback processing obviously depend on past data. Even for simple sampling, live data exists in the working memory during idle periods since buffering is a great help to reduce data transmit frequency to backend systems.

To reduce the static energy of the working memory while preserving the live data, there exist two kinds of solutions, one is to allow the working memory to enter a low power mode and the other is data migration before power gating. Based on the former idea, there exists a wide variety of power saving techniques for working memory while preserving data. These researches trade between static power and wake-up latency [3, 6, 10]. However, the static power reduction is limited and remained static power is not negligible for long idle periods.

In this section, we focus on the migration of live data to a new generation non-volatile memory (NVM) [9]. As a traditional non-volatile memory, FLASH is popular and widely used as storage rather than memory due to its huge access energy, slow access speed and limited endurance. On the other hand, new generation NVMs have successfully solved these problems [4, 7] and already become available [1, 2]. Therefore, we revisit data-aware power management with the new NVMs. In case that such non-volatile memory is used for those real-time systems, further power reduction can be expected if coupled with intelligent power management.

Firstly, we classify sleep modes of micro-controllers (having working memory internally) into two. One is **shallow sleep**, which can preserve data. The other is **deep sleep**, which loses data but consumes no static power. If we only consider shallow sleep modes, there is a trade-off between PG overhead energy and reduced static power. i.e. sleep mode that can reduce more static power requires larger PG overhead energy. Based on this trade-off, an optimal power management is determined by BET, which can be calculated as follows.

$$BET = \frac{PG\ overhead\ energy}{Static\ power\ reduction\ by\ PG}$$

Here, each shallow sleep mode has a different BET since its PG overhead energy and static power reduction are given by its differences from the shallower sleep mode. When the length of idle time is given, the optimal sleep mode is the one with the longest BET that is also shorter than the length of the idle time. When deep sleep is chosen, the situation is more difficult. Since, all live data should be migrated to NVM before PG and restored after PG.

For optimal power management with a combination of shallow and deep sleep, data migration overhead energy have to be considered in addition to PG overhead energy. Thus, an extended BET (eBET) is theoretically given by

$$\begin{aligned} eBET &= \frac{Overhead\ energy\ of\ Deep\ sleep}{Static\ power\ reduction\ by\ Deep\ sleep} \\ &= \frac{PG\ overhead\ energy + Data\ migration\ Overhead}{Static\ power\ reduction\ by\ Deep\ sleep}. \end{aligned}$$

Here, PG overhead energy and static power reduction are given by the differences between itself and the deepest shallow sleep mode.

The data migration overhead energy depends on access energy of NVM. If we use FLASH, the overhead is huge and eBET can be hours. Then deep sleep can never be chosen. On the other hand, with new generation NVM, eBET is in the order of 10 to 100 ms and deep sleep becomes available. Therefore, in this section, we focus on deep sleep with the new generation NVM.

An essential difference between these two overheads is that the PG overhead energy always depends on only one idle period but the data migration overhead energy may depend on multiple idle periods since migrated data may not be needed immediately after an idle period ends and it may span several idle periods. Therefore, eBET and the optimal solution cannot determine by any simple guideline, unlike the traditional BET. It is necessary to consider both the lengths of idle periods and the amount of live data together.

Moreover, modern embedded systems run multiple tasks which have different characteristics. Then, to find the optimal power management for each idle period is quite difficult. For example, execution intervals of two tasks are different, a task may wake up when another task is idle. As a result, the idle periods are divided by each of these independent tasks. The lifetimes of the live data in the same memory are mixed.

To cope with this problem, our approach is as follows. Firstly, we clarify all parameters that affect the optimal power management with NVM and formulate an optimization problem with these parameters. Then we propose a rapid solution, since a brute force approach requires exponential computation cost. The proposed solution can find a near optimal solution in polynomial time. Finally, we evaluate our approach with real hardware and a simulator. We collect the parameters, then compare speed and accuracy of the obtained solution with a brute force approach. Finally, the energy reduction ratio is evaluated.

The primary contributions of this section are as follows.

- We introduce a power gating method that is applied to a working memory of micro-controllers.
- We propose a power management that migrates live data between volatile and non-volatile memories to avoid data loss.
- We develop a rapid algorithm to find sleep modes and deciding whether to migrate data or not appropriately.

5.6.2 Motivating Example

As we mentioned, multiple tasks run in the system and their executions overlap. Thus, the optimal power management may not be intuitive.

A motivating example is shown in Fig. 5.35. In this example, the x-axes represent time and the y-axes indicate power consumptions. Three charts correspond to different power managements. The gray boxes represent periodic processes. Note that the dynamic energy is the same regardless of the power management of idle periods and omitted from the discussion in this section. Idle periods exist between processes. The blue hatched boxes represent the static energy of the shallow sleep, which is consumed in idle periods. In this example all of the static energy is 10. When the deep sleep is applied, the static energy becomes 0. More precisely, the blue hatched boxes represent the difference between the reduced static energy and the PG overhead energy in the idle period.

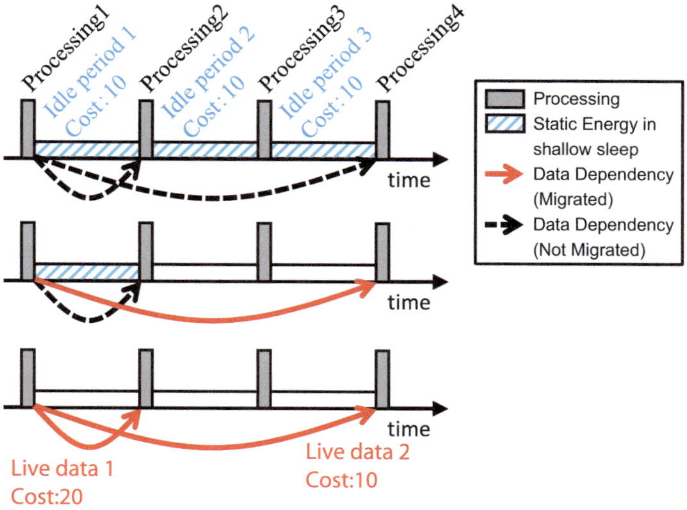

Fig. 5.35 Motivating example

The arrows represent data dependency. For example, data 1 is generated in processing 1 and consumed in processing 2, data 2 is also generated in processing 1 but consumed in processing 4. Thus, data 1 is alive in idle period 1 and data 2 is alive in idle period 1, 2 and 3. The migration energy of data 1 and data 2 are 20 and 10 respectively. This means that these data come from different kinds of sensors and the size of data 1 is twice of data 2.

The uppermost chart represents a power management that always uses the shallow sleep. Since the shallow sleep can preserve the live data, no migration overhead is needed. Thus, the total energy overhead becomes $10 + 10 + 10 = 30$.

In the same manner, the lowermost chart represents a power management that always uses the deep sleep, which can cut the static power completely. On the other hand, data migrations are required to preserve the live data. Thus, the total energy overhead becomes $20 + 10 = 30$.

Finally, the central chart represents a power management that uses the shallow sleep in idle period 1 and the deep sleep in idle periods 2 and 3. Only data 2 is migrated. Since data 1 is not migrated, the shallow sleep must be used and static energy is consumed. Thus, the total energy overhead becomes $10 + 10 = 20$.

In this section, we focus on the optimal power management that requires a combination of the shallow and the deep sleep modes properly. In this example, there are only 3 idle periods and 2 data. However, it is not easy to extend to solve a general problem that can cover real applications. To cope with this problem, we model this and propose a rapid solution.

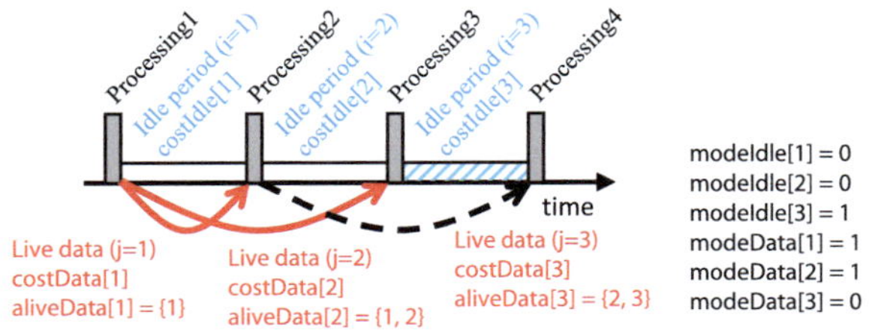

Fig. 5.36 Parameter description

5.6.3 *Data-Aware Power Management*

In this section, we firstly identify input and output parameters that describe our target system. Then, we explain our rapid solution, which minimizes the power consumption during idle periods.

5.6.3.1 Problem Formulation

An example showing all variables is presented in Fig. 5.36. This example corresponds to Fig. 5.35 in the previous section. For each variable, i $(i = 1, 2, \ldots, N)$ represents ID of the idle period and j $(j = 1, 2, \ldots, M)$ represents ID of the live data.

Reduced energy in idle period i ($costIdle[i]$) is given by

$$costIdle[i] = idleTime_i \times (SP_{shallow} - SP_{deep}) + EOH_{deep} - EOH_{shallow}. \quad (5.3)$$

Here $idleTime_i$ represents the length of idle period i, $SP_{shallow}$ and SP_{deep} represent static power during shallow and deep sleep modes respectively, $EOH_{shallow}$ and EOH_{deep} represent PG overhead energy of shallow and deep sleep respectively. Note that if the length of the idle period is shorter than BET, $costIdle[i]$ is less than 0. $modeIdle[i]$ is a binary variable and represents sleep mode in idle period i.

The parameters that related to live data are $costData[j]$, $aliveData[j]$ and $modeData[j]$. $costData[j]$ represents migration overhead energy of live data j, which includes write and read energy of micro-controller and non-volatile memory. Migration overhead energy of live data j is given by

$$costData[j] = dataSize_j \times (WE + RE). \quad (5.4)$$

Here $dataSize_j$ represents the amount of live data j, which depends on the type of sensor, WE and RE represent consumed energy per byte of write and read operations respectively. $aliveData[j]$ represents data dependencies between processing and is explained by an ID set of idle periods in which live data j is alive. For example, in Fig. 5.36, the live data 2 is alive in the idle period 1 and 2, so $aliveData[2] = \{1, 2\}$.

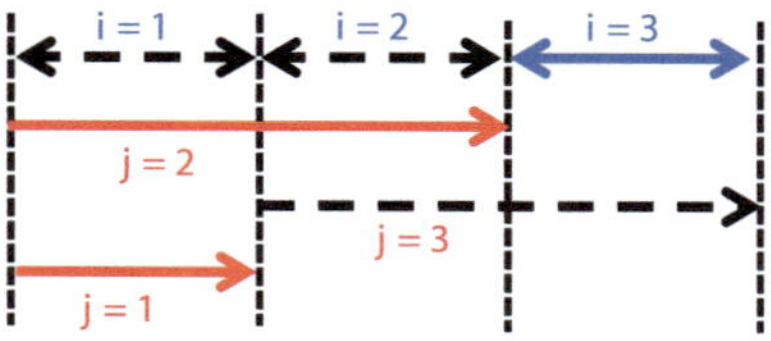

Fig. 5.37 Expression for algorithm explanation

modeData$[i]$ is a binary variable and represents whether live data j is migrated or not.

Using these parameters, we formulate a problem as shown below.

$$\min \sum_i modeIdle[i] \times costIdle[i] + \sum_j modeData[j] \times costData[j] \quad (5.5)$$

$$\begin{aligned} \text{s.t. } & modeIdle[i] = 0 \Rightarrow \forall j \;\; aliveData[j] \not\ni i \text{ or } modeData[j] = 1 \\ & \qquad\qquad (i = 1, \ldots, N) \qquad (5.6) \\ & modeIdle[i] \in \{0, 1\},\; modeData[j] \in \{0, 1\} \end{aligned}$$

The first clause of the object function (5.5) is the total energy that is consumed in all idle periods. The second clause of the object function (5.5) is the sum of the data migration overheads. The constraint condition (5.6) means that in order to use deep sleep in an idle period, all live data that are alive in the idle period must be migrated.

5.6.3.2 Solution

Since all variables are binary, we can solve the formulated problem by brute force approach. However it requires exponential computation complexity, so we might not find an optimal solution in real time, even though it can be done offline. To cope with this problem, we propose a rapid solution that finds a near optimal solution rapidly by considering idle time and the amount of live data.

As an explanation, we use an example shown in Fig. 5.37, which is generated from Fig. 5.36. Lines with double-headed arrows represent idle periods; of which the ones having solid blue lines are those with shallow sleep applied while the ones having dotted black lines are those with deep sleep applied. Lines with single-headed arrows represent live; solid red lines mean live data is migrated while dotted black lines are used for no live data migration. The single-headed arrows represent live data; the red and black broken lines mean that the live data is migrated and not migrated respectively. The length of the single-headed arrow represents the lifetime.
Overview of the rapid algorithm
In our approach, we use dynamic programming. The two key observations are (1)

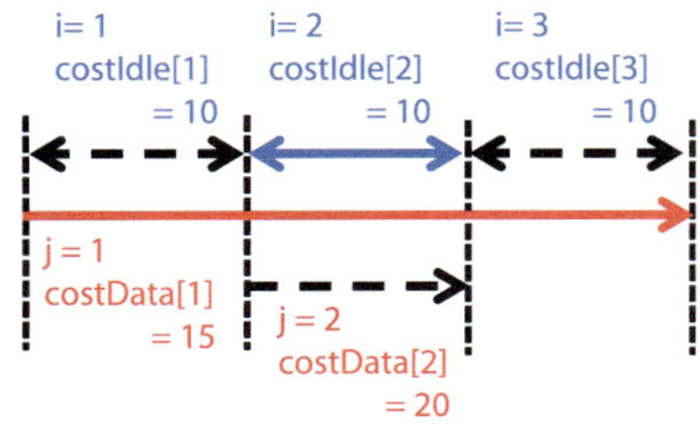

Fig. 5.38 Simple example

lifetimes of live data may overlap and (2) the migration overhead energy of live data may exceed energy reduction.

We explain these observations one by one with Fig. 5.38. In the first case, we can see that the lifetimes of live data 1 and 2 are overlapped in idle period 2. Assuming that live data 1 is not migrated, migration of live data 2 makes no sense because deep sleep cannot be used in idle period 2. So we need not search for the case that live data 1 is not migrated and live data 2 is.

In the second case, we can see that migration overhead energy of live data 2 (20) is larger than the energy reduction in idle period 2 if deep sleep is applied. It means that migrating live data 2 surely waste more energy and this is true regardless of other live data. So we can determine that live data 2 is not migrated.

Inclusion Property of Live Data

In our algorithm, we firstly construct inclusion property of live data. If there are two live data j_c and j_p that meet the condition below, we assume live data j_c as a child of live data j_p.

- The lifetime of live data j_p is a superset of that of live data j_c.
- There is no live data whose lifetime is superset of that of live data j_c and is subset of that of live data j_p.

Here, the first condition is defined below.

$$\begin{aligned} &aliveData[j_p] \supset aliveData[j_c] \\ &or \\ &aliveData[j_p] = aliveData[j_c] \;\; and \;\; j_p \leq j_c \end{aligned} \tag{5.7}$$

The first line means that the lifetime of live data j_p is a superset of that of live data j_c (such that the two lifetimes are not the same), the third line means that the two lifetimes are the same and ID j_p is larger than j_c. Here the ID set of children of live data j is represented as *children*$[j]$. Also, the ID set of descendants of live data j is represented as *descendants*$[j]$.

Rapid Algorithm

After generating the inclusion property of live data, it is ready to use algorithm 1. We start our algorithm from a state that deep sleep is applied to all idle periods and all live data are migrated, so 0 is set to *modeIdle*$[i]$ and 1 is set to *modeData*$[j]$ as an initial state. First, algorithm 1 should be applied to the root node. Here if there are several root nodes, algorithm 1 should be applied to each root node. Then it is applied to every node through recursive calls.

Algorithm 1. $select\_sleepMode(j_{input})$

```
Input: costIdle[i], costData[j], aliveData[j], j_input
Output: modeIdle[i], modeData[j]
1: migEnergy ← 0
2: for j ∈ children(j_input) do
3:   if modeData[j] = 1 then
4:     migEnergy ← migEnergy + select_sleepMode(j)
5:   end if
6: end for
7: migEnergy ← migEnergy + costData[j_input]
8: statEnergy ← 0
9: for i ∈ aliveData[j_input] do
10:   if modeIdle[i] = 0 then
11:     statEnergy ← statEnergy + costIdle[i]
12:   end if
13: end for
14: if statEnergy ≥ migEnergy then
15:   return migEnergy
16: else
17:   for i ∈ aliveData[j_input] do
18:     modeIdle[i] ← 1
19:   end for
20:   for j ∈ descendants(j_input) do
21:     modeData[j] ← 0
22:   end for
23:   return 0
24: end if
```

In lines 1–7, *migEnergy* is calculated, which is the reduced energy if live data j_{input} is not migrated. In that situation, all children live data of live data j_{input} should not be migrated (see the first case of Sect. 5.6.3.2). In lines 8–13, *statEnergy* is calculated, which is the cumulated energy if shallow sleep is applied in idle periods during which live data j_{input} is alive. In lines 14–24, *migEnergy* and *statEnergy* are compared. If $statEnergy \geq migEnergy$, it means that migrating live data j_{input} might reduce energy (depending on other live data). Otherwise, it means that live data j_{input} should not be migrated (regardless of other live data, see the second case of Sect. 5.6.3.2).

5.7 Conclusion

In this chapter, we introduced detailed power management techniques and how to realize normally-off computing in each component.

Computer systems consist of some of these components. Therefore, to realize normally-off computing systems, these components are carefully combined. Three practical normally-off computer systems will be introduced in the next chapter.

References

1. FRAM series datasheet. http://www.tij.co.jp/jp/lit/ds/slas639h/slas639h.pdf
2. MN101L series datasheet. http://www.semicon.panasonic.co.jp/ds8/c2/21705-015.pdf
3. Agarwal, K., Nowka, K., Deogun, H., Sylvester, D.: Power gating with multiple sleep modes. In: Proceedings of the 7th International Symposium on Quality Electronic Design, pp. 633–637. IEEE Computer Society (2006)
4. Burr, G., Kurdi, B., Scott, J., Lam, C., Gopalakrishnan, K., Shenoy, R.: Overview of candidate device technologies for storage-class memory. IBM J. Res. Dev. **52**(4.5), 449–464 (2008). doi:10.1147/rd.524.0449
5. Hu, Z., Buyuktosunoglu, A., Srinivasan, V., Zyuban, V., Jacobson, H., Bose, P.: Microarchitectural techniques for power gating of execution units. In: International Symposium on Low Power Electronics and Design, pp. 32–37 (2004). doi:10.1109/LPE.2004.1349303
6. Kim, S., Kosonocky, S.V., Knebel, D.R., Stawiasz, K.: Experimental measurement of a novel power gating structure with intermediate power saving mode. In: Proceedings of the 2004 International Symposium on Low Power Electronics and Design, pp. 20–25. IEEE (2004)
7. Kryder, M.H., Kim, C.S.: After hard drives? what comes next? IEEE Trans. Magn. **45**(10), 3406–3413 (2009)
8. Nakada, T., Okamoto, K., Komoda, T., Miwa, S., Sato, Y., Ueki, H., Hayashikoshi, M., Shimizu, T., Nakamura, H.: Design aid of multi-core embedded systems with energy model. IPSJ Trans. Adv. Comput. Syst. **7**(ACS46), 37–46 (2014)
9. Nakada, T., Shigematsu, T., Komoda, T., Miwa, S., Sato, Y., Ueki, H., Hayashikoshi, M., Shimizu, T., Nakamura, H.: Data-aware power management for periodic real-time systems with non-volatile memory. In: 3rd IEEE Nonvolatile Memory Systems and Applications Symposium (NVMSA), 6 p (2014)
10. Pakbaznia, E., Pedram, M.: Design and application of multimodal power gating structures. In: 2009 10th International Symposium on Quality Electronic Design, pp. 120–126 (2009). doi:10.1109/ISQED.2009.4810281
11. Pollack, F.: Micro32 conference keynote. Intel Corporation (1999)
12. Puri, R., Stok, L., Bhattacharya, S.: Keeping hot chips cool. In: 42nd Design Automation Conference, pp. 285–288 (2005). doi:10.1109/DAC.2005.193818
13. Yoda, H.: Progress of STT-MRAM technology and the effect on normally-off computing systems. In: Session 11.3, Technical Digest of International Electron Devices Meeting (IEDM) (2012)
14. Kitagawa, E., Fujita, S., Nomura, K., Noguchi, H., Abe, K., Shimomura, N., Ito, J., Yoda, H., Daibou, T., Kato, Y., Kamata, C., Kashiwada, S.: Impact of ultra low power and fast write operation of advance perpendicular MTJ on power reduction for high performance mobile CPU. In: Session 29.4, Technical Digest of International Electron Devices Meeting (IEDM) (2012)
15. Abe, K., Fujita, S., Lee, T.H.: Architecture of three-dimensional circuit using nanoscale memory devices. European Micro and Nano Systems (EMN04), pp. 225-229, Oct. 2004
16. Abe, K., Fujita, S., Lee, T.H.: Novel nonvolatile logic circuits with three-dimensionally stacked nanoscale memory device. Proceedings of the 2005 NSTI Nanotechnology Conference **3**, 203–206 (2005)
17. Abe, K., Nomura, K., Ikegawa, S., Kishi, T., Yoda, H., Fujita, S.: Hierarchical nonvolatile memory with perpendicular magnetic tunnel junctions for normally-off computing. In: The

2010 International Conference on Solid State Devices and Materials (SSDM), Tokyo, pp. 1144-1145, Sep. 2010
18. Yamamoto, S., Sugahara, S.: Nonvolatile static random access memory using magnetic tunnel junctions with current-induced magnetization switching architecture. Jpn. J. Appl. Phys. **48**(2009), 043001 (2009)
19. Ohsawa, T., Koike, H., Miura, S., et al.: 1Mb 4T-2MTJ nonvolatile STT-RAM for embedded memories using 32b fine-grained power gating technique with 1.0ns/200ps wake-up/power-off times. In: Symposium VLSI Circuits, pp. 46–47, July 2012
20. Abe, K., et al.: Novel hybrid DRAM/MRAM design for reducing power of high performance mobile CPU. In: IEDM Technical Dig., pp. 243–246 (2012)
21. Kawasumi, A., et al.: Circuit techniques in realizing voltage-generator-less STT MRAM suitable for normally-off-type non-volatile L2 cache memory. In: 5th IEEE International Memory Workshop (IMW), pp. 76–79 (2013)
22. Noguchi, H., et al.: A 250-MHz 256b-I/O 1-Mb STT-MRAM with advanced perpendicular MTJ based dual cell for nonvolatile magnetic caches to reduce active power of processors. In: VLSI Technology Symposium, C108 - C109 (2013)
23. Tanaka, C., et al.: Normally-off type nonvolatile SRAM with perpendicular STT-MRAM cells and smallest number of transistors. In: International Conference on Solid State Devices and Materials (SSDM), pp. 1092–1093 (2013)
24. Noguchi, H., et al.: A 3.3ns-Access-Time 71.2uW/MHz 1Mb embedded STT-MRAM using physically eliminated read-disturb scheme and normally-off memory architecture. In: ISSCC (2015)
25. The gem5 simulator: http://gem5.org/
26. SPEC CPU2006 benchmark suite: http://www.spec.org/
27. Ando, K., Ikegawa, S., Abe, K., Fujita, S., Yoda, H.: Normally-off computer: new roles of nonvolatile devices in future computer systems. In: Sustainable Green Computing. IGI Press, ISBN 978-1-4666-1842-8, June, 2012
28. Takeda, S., et al.: Low-power cache memory with state-of-the-art STT-MRAM for high performance processors. In: ISOCC (2015)
29. Kitagawa, E., et al.: Sub-30nm p-MTJ with small switching current, large MR, and high thermal stability. In: 12th Joint MMM/Intermag Conference (2013)

Chapter 6
Research and Development of Normally-Off Computing—NEDO Project

Takashi Nakada, Shinobu Fujita, Masanori Hayashikoshi, Shintaro Izumi, Yoshikazu Fujimori and Hiroshi Nakamura

Abstract Based on normally-off computing design methodology, we developed three practical systems that introduced normally-off computing. These systems are healthcare, mobile information device and sensor node for social infrastructure. These systems are selected from different types of application areas. Through implementing a variety of systems, universalness of normally-off computing is confirmed. Also we introduce our "Normally-Off Computing Project", which supports these practical developments.

Keywords Normally-off computing project · Healthcare · Mobile information device · Sensor node

T. Nakada (✉) · H. Nakamura
The University of Tokyo, Bunkyo, Tokyo, Japan
e-mail: nakada@hal.ipc.i.u-tokyo.ac.jp

H. Nakamura
e-mail: nakamura@hal.ipc.i.u-tokyo.ac.jp

S. Fujita
Toshiba Corporation, Minato, Tokyo, Japan
e-mail: shinobu.fujita@toshiba.co.jp

M. Hayashikoshi
Renesas Electronics Corporation, Koto, Tokyo, Japan
e-mail: masanori.hayashikoshi.cj@renesas.com

S. Izumi
Kobe University, Kobe, Hyogo, Japan
e-mail: shin@cs28.cs.kobe-u.ac.jp

Y. Fujimori
ROHM Co., Ltd., Kyoto, Kyoto, Japan
e-mail: yoshikazu.fujimori@mnf.rohm.co.jp

T. Nakada and H. Nakamura (eds.), *Normally-Off Computing*,
DOI 10.1007/978-4-431-56505-5_6

6.1 Overview

In previous chapters, we explained low power technologies and normally-off computing design methodology. In Chap. 4, we introduced device technologies that realize small BET, architecture technologies that adapt wide variety of requirement, and software technologies that manage dynamic behavior. In Chap. 5, details of above technologies are explained.

In this chapter, we describe three practical normally-off systems that are being developed in our "Normally-Off Computing Project" [1] supported by NEDO (New Energy and Industrial Technology Development Organization)/METI (Ministry of Economy, Trade and Industry) in Japan. This is a 5 year project started in 2011. Toshiba Corporation, Renesas Electronics Corporation, ROHM Corporation, and the University of Tokyo join this project (Fig. 6.1). All the industrial members have their own strong application fields and are responsible for developing world-leading normally-off computing in their fields. The University of Tokyo is working with them to put all their knowledge together and establish general normally-off computing design methodology for forthcoming comfortable and power efficient information society.

Our "Normally-Off Computing Project" aims to achieve higher performance per power for wide varieties of computing systems including the next generation sensor networks, mobile devices, servers and other equipment. Its strategy is to make full use of non-volatile memory by cooperative development of hardware and software.

The granularity depends on the characteristic of memory access required by applications. Thus, the solutions also depend on the target fields or applications.

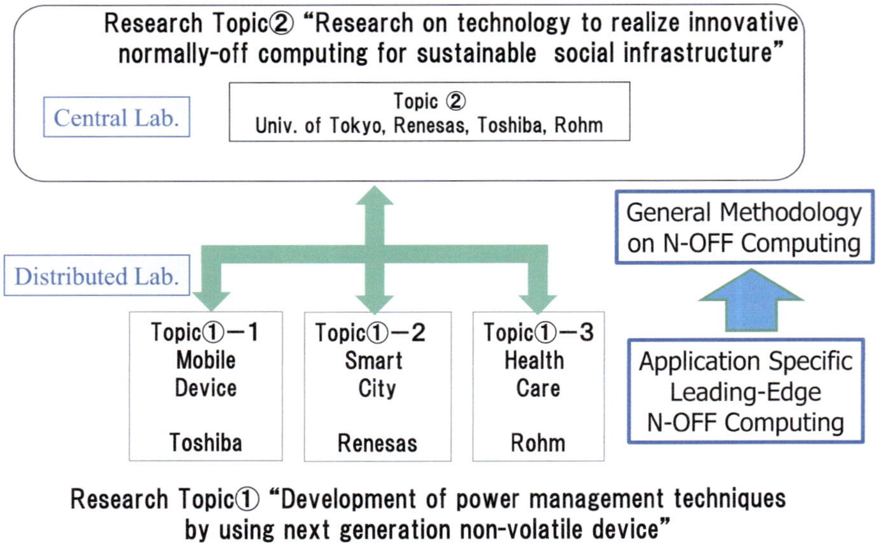

Fig. 6.1 Organization of normally-off computing project

The goal of program (1) is to realize drastic power reduction dedicated for each application field. This program is performed by three industrial companies, which are Toshiba Corporation, Renesas Electronics Corporation, and Rohm Corporation, independently. We call the laboratory engaged in program (1) distributed laboratory.

The goal of program (2) is to investigate and develop generalized normally-off computing, which can be applied widely to our future sustainable and sophisticated information society. Its goal is development of computer platform independent from specific applications rather than application specific appliances. This program is carried out in central laboratory located in the University of Tokyo. Researchers from the companies also join this laboratory.

The distributed and the central laboratories tightly collaborates each other through generalization and specialization feedback loops. We are trying to develop novel and efficient normally-off computing through these feedback loops.

For the success of program (2), the development of generalized normally-off computing, it would be better to have experience and insight of research and development on wider varieties of applications. Then, the following question arises; how many applications are required? Before starting this project, we summarized what kinds of application characteristics significantly affect the development of normally-off computing.

Applications are categorized according to two characteristics, "peak performance" and "workload regularity" as shown in Fig. 6.2. We picked up these two characteristics due to the following reasons. First, peak performance largely affects system configuration and memory hierarchy. Secondly, temporal granularity of activity is significantly affected by workload regularity.

Then we selected three application fields, mobile device, smart city and health care, because these applications cover a wide area of the whole application field as shown in from Fig. 6.2. We introduce how to optimize them with our low power technologies, which are introduced in previous chapter.

The mobile devices require high peak performance to provide intelligent data processing. Some applications have steady workload such as encoding or decoding,

Fig. 6.2 Target applications

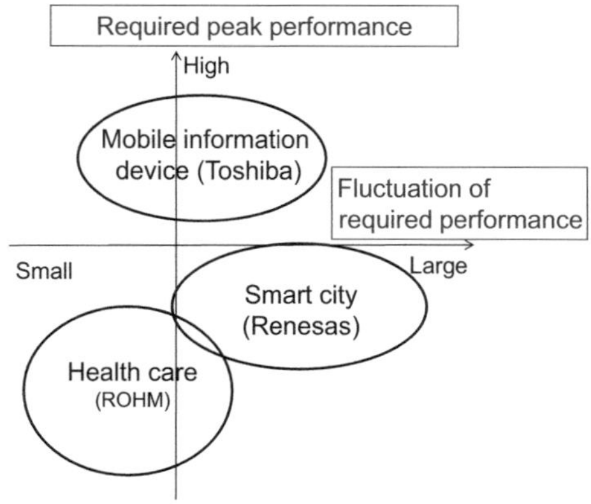

but other applications have irregular workload such as web browsers or games. With these devices, the energy consumption of cache memories is not negligible. Thus, energy efficient cache technologies, which are introduced in Chap. 5, Sect. 6.3.

In the smart city, there are lots of smart sensors. In most cases, these sensors periodically collect data. However once unusual situation is detected, additional workload may be invoked. Thus we categorize it into irregular workload and their task scheduling and adaptive power management, which are introduced in Sects. 5.4 and 5.5, are important.

Finally, the health care systems also consist of many sensors, but its application requires long term data collection and fixed statistical and periodical processing. To minimize their energy consumption, nonvolatile logic circuit technologies, which are introduced in Sect. 5.2, are important.

6.2 Healthcare

6.2.1 Background

Because of the advent of an aging society in Japan, mobile health plays an ever more prominent role [2]. Daily-life monitoring is especially important in preventing lifestyle diseases, which have rapidly increased the number of patients and elderly people requiring nursing care. Our goal is the monitoring and display of vital signals and physical activity in daily life to improve users' quality of life and realize a smart society.

This report specifically describes a wearable biosignal monitoring system, which can acquire long-term instantaneous heart rate (IHR) data and an acceleration value. The physical activities in daily life (e.g., locomotive, household activities) are classifiable using a triaxial accelerometer [3]. The IHR, which is calculated from the interval of R-waves in electrocardiogram (ECG), is useful for heart disease detection, heart rate variation (HRV) analysis [4], and exercise intensity estimation [5].

The key factors affecting wearable system usability are miniaturization and weight reduction. Battery mass and power consumption must be reduced because battery mass dominates wearable systems. To reduce the power consumption, a wearable and wireless ECG telemetry system [6, 7] and single-chip ECG monitoring system LSIs [8–10] have been developed. However, strict limitations on power consumption and electrode distance of wearable ECG monitors render them sensitive to noise of various kinds. Especially, if a subject is not at rest (e.g. during exercise), the signal-to-noise ratio (SNR) of ECG signals will be significantly degraded.

To realize the low-power and noise-tolerant system, we proposed an ECG-SoC using normally-off architecture and robust IHR monitor.

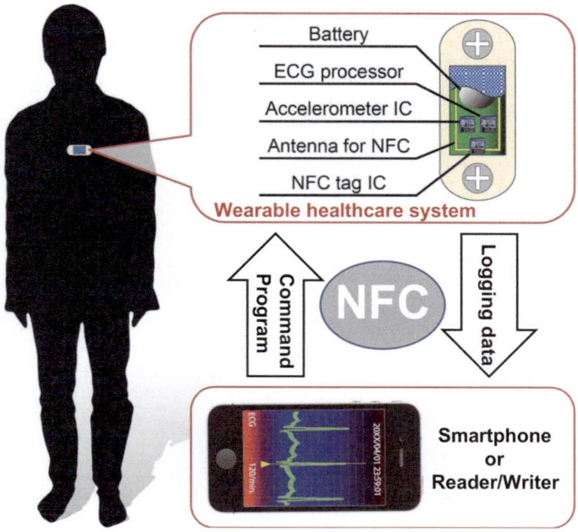

Fig. 6.3 Wearable healthcare system overview

6.2.2 System Description and Normally-Off ECG-SoC Architecture

Figure 6.3 presents an overview of the wearable healthcare system, comprising the proposed ECG-SoC, Near Field Communication (NFC) tag IC, and accelerometer IC. The NFC is used for program loading, individual optimization, and data retrieval from the ECG-SoC.

The proposed ECG-SoC consists of an ECG sensing block, NVMCU, and extra interfaces (see Fig. 6.4). The ECG sensing block has an analog front end (AFE), an 8-bit SAR ADC, and a robust IHR extractor. The NVMCU (see Fig. 6.5) includes a Cortex M0 (CM0) core with ferroelectric-based nonvolatile FFs (NVFF) [11], a 16 Kb 6T-4C NVRAM for instruction and data memory, and peripherals. Because the frequency range of vital signals is low, both the standby power reduction and sleep time maximization is important to system level power reduction.

We employed a 6T-4C memory cell (see Fig. 6.6a) for the NVRAM because it has the advantage of non-volatility and fast access time [12]. However, active power dissipation is caused by the large ferroelectric capacitor, which is connected directly to plate-lines and internal nodes. This work presents plate-line charge-share and bit-line non-precharge techniques to reduce the power overhead (see Fig. 6.6). Additional switches between plate-lines are used to share the charge used for store and recall operations. The charge is transported sequentially from a plate-line to the next one. Furthermore, the 6T-4C cell is tolerant to the half-select problem because the ferroelectric capacitors are connected to the internal nodes. Therefore, the bit-line precharge can be omitted and only equalizing is used in this design. As shown in Fig. 6.7, energy consumptions of NVRAM in store, recall, write, and read operations

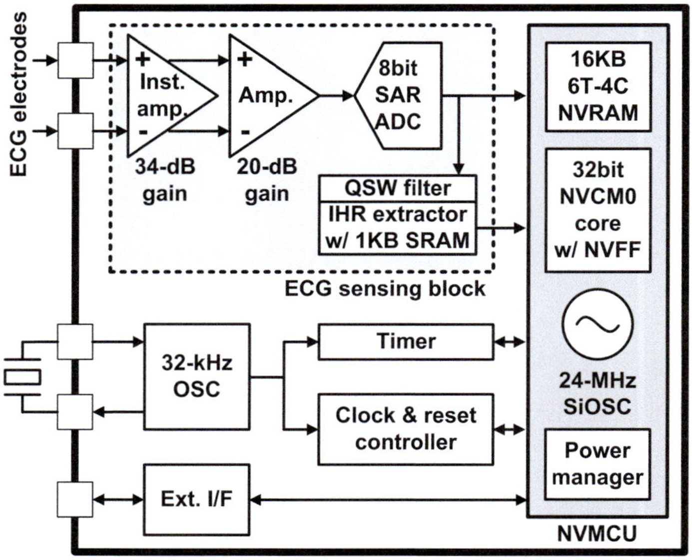

Fig. 6.4 Block diagram of normally-off ECG-SoC

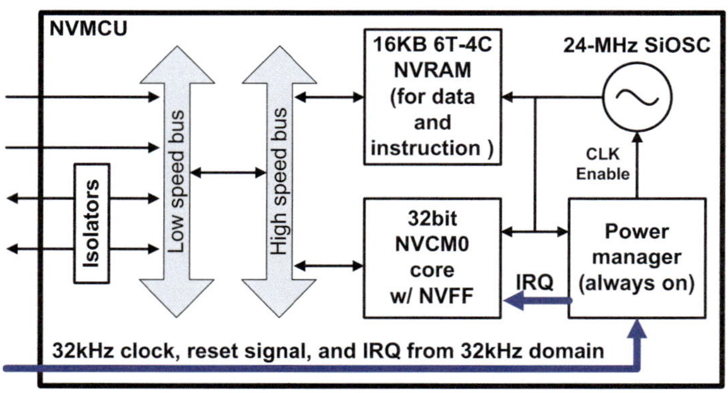

Fig. 6.5 Block diagram of nonvolatile MCU (NVMCU)

are reduced respectively by 22, 11, 74, and 77% compared with original 6T-4C memory.

Figure 6.8 shows the shut-down and wake-up sequence of NVMCU. The operating frequency of the NVMCU is 24 MHz, whereas the operating frequency of other digital blocks is 32 kHz. Slow signals in the 32-kHz domain are synchronized at the low-speed bus to the 24-MHz domain. Standby current of the entire 24-MHz domain including an on-chip 24-MHz oscillator can be cut when the state of CM0 core transits to deep sleep. Then the data in the NVRAM and register values of CM0 core in the NVFF are stored sequentially to ferroelectric capacitors. The data and register values of NVMCU will be recalled if the interrupt occurs from the 32 kHz

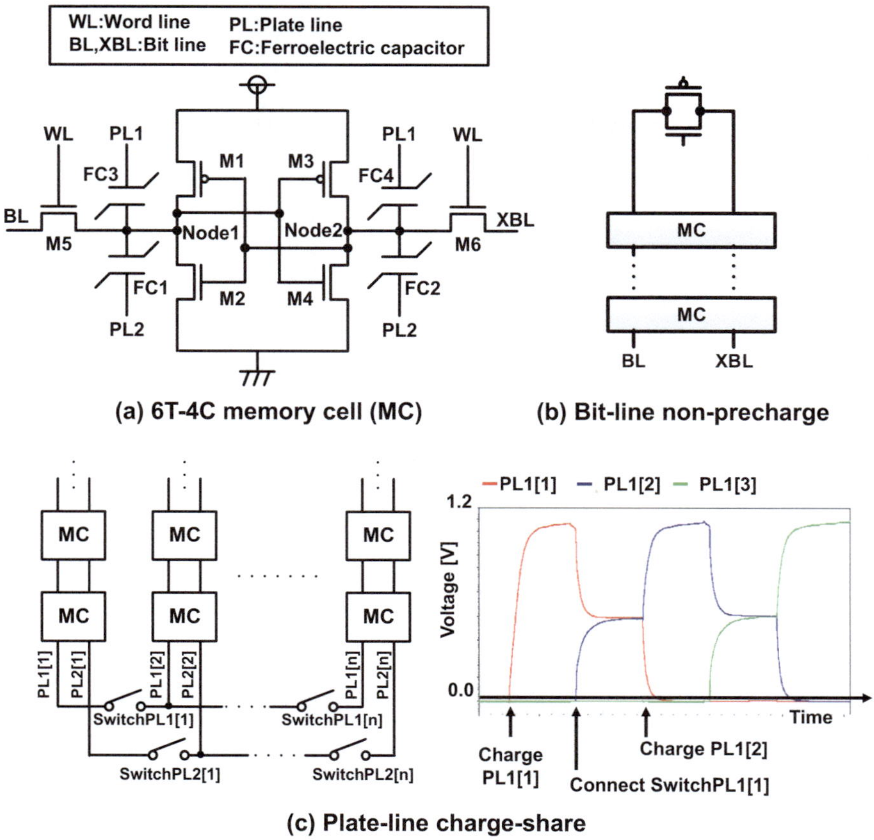

Fig. 6.6 Proposed NVRAM with **a** 6T-4C memory cell, **b** bit-line non-precharge technique, and **c** plate-line charge-share technique

domain. The store and recall operation dissipate up to 25-us overhead. As shown in Fig. 6.9, the energy break-even time using the normally-off function is about 0.5% active ratio.

6.2.3 Noise Tolerant and Low-Power ECG Sensing Block

The ECG sensing block has an analog front end (AFE), a 12-bit SAR ADC, and a robust IHR monitor. Although the IHR monitor using high-order eight-bit data of the ADC, 12-bit ADC output also directly connected to Cortex M0. The AFE includes a 34-dB gain instrumental amplifier and a 20-dB gain amplifier as shown in Fig. 6.10. The ADC sampling rate can be set to 1 kSamples/s for ECG processing

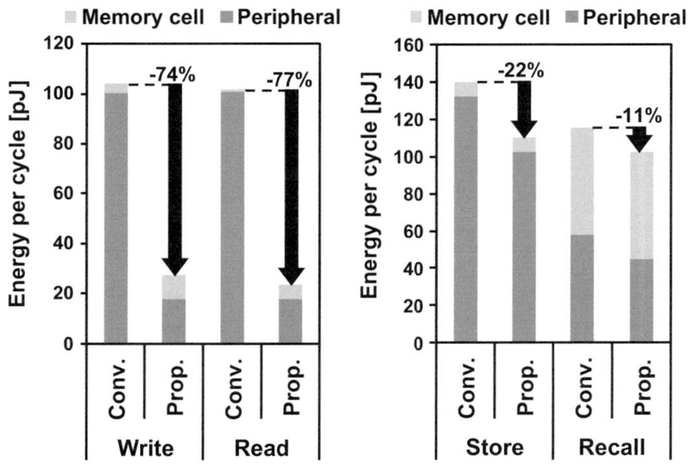

Fig. 6.7 Measured energy consumption of conventional and proposed 6T-4C NVRAM

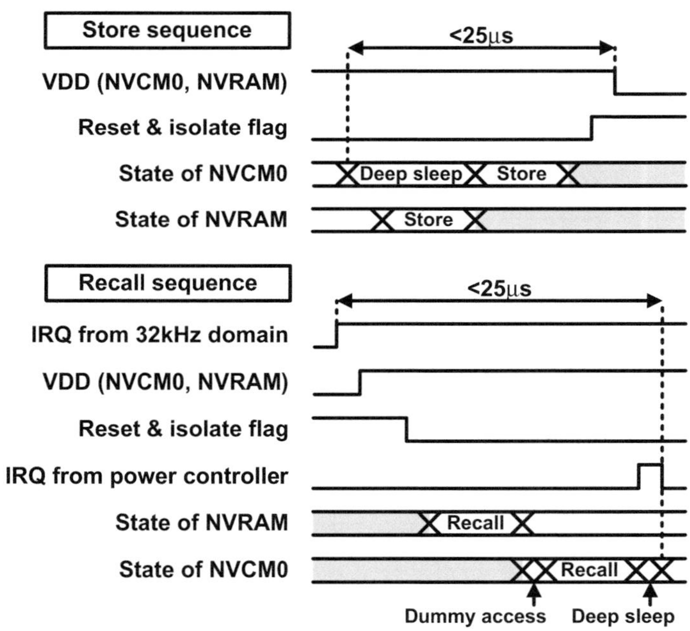

Fig. 6.8 Timing diagram of store/recall sequence in NVMCU

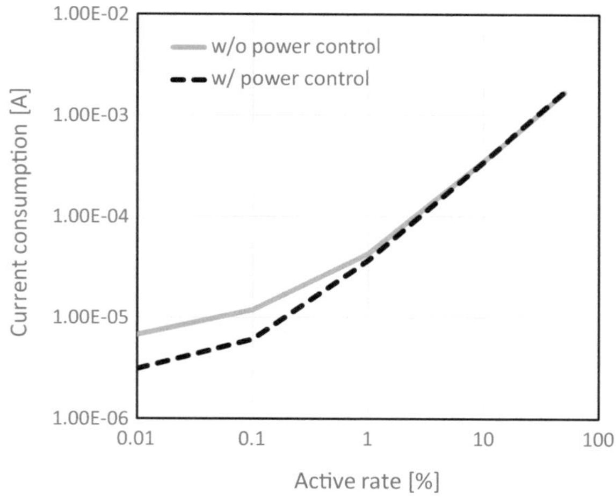

Fig. 6.9 Measured current consumption of NVMCU

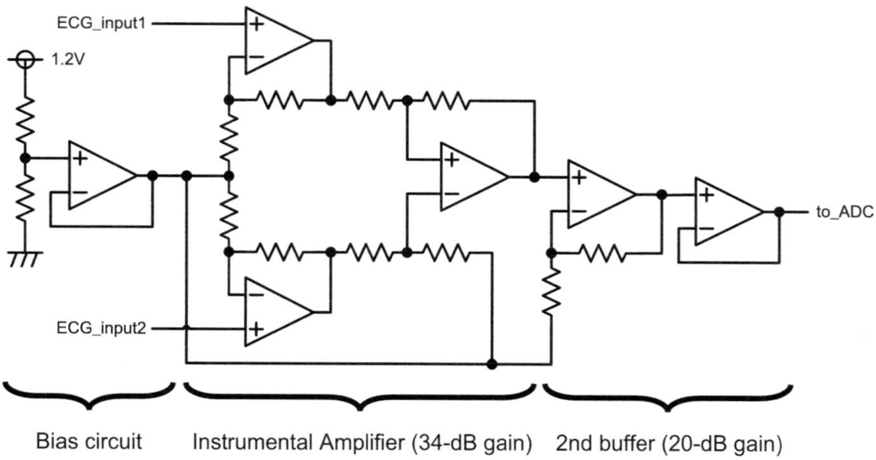

Fig. 6.10 Analog front-end circuits of ECG sensing block

mode and 128 Samples/s for IHR monitoring mode. The robust IHR monitor is the main contribution of this study.

The ECG signal in wearable systems is sensitive to various noises because the electrode distance, size and the battery capacity are strictly limited. The SNR will be especially degraded if a user is not at rest. The purpose of our approach is digital signal processing to mitigate the performance requirements for the analog portion and to minimize the overall system power consumption. We implemented a quadratic spline wavelet (QSW) filter and a two-stage IHR extractor. The QSW is commonly used as the low-power noise reduction method for ECG [13]. The hum-noise and baseline wander are well suppressed using QSW. However, it is difficult to remove these noises only using QSW because it has similar frequency range of the QRS

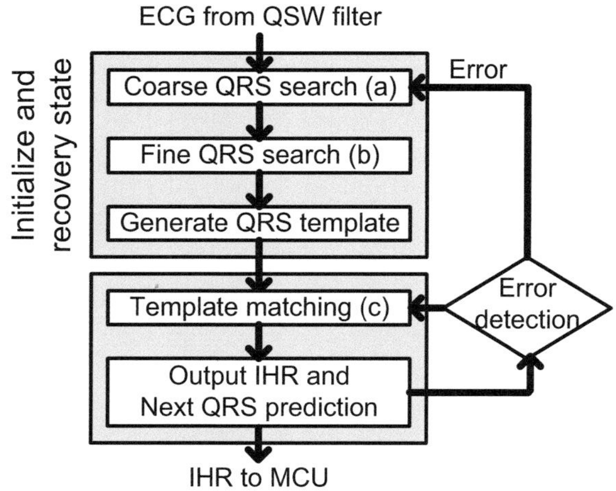

Fig. 6.11 Flow chart of IHR extractor

complex. Figures 6.11 and 6.12 show the algorithm of the IHR extractor. In the first stage, the template data of QRS complex are generated autonomously using the extended version of a coarse-fine short-term autocorrelation (STAC) [14]. Next, template matching is conducted to extract QRS complexes. Whenever the QRS complex is extracted, the IHR and the QRS template are updated by adding the detected QRS complex to the previous template. Then, the time at which the next QRS complex occurs is predicted from the beat-to-beat variation.

The prediction result is used to maximize the sleep time of the ADC and IHR extractor. Even if misdetection or false detection occurs because of arrhythmia or intense noise, the IHR extractor can be awake and recover the error because the coefficient of autocorrelation will decrease rapidly when such an error occurs.

The IHR extractor can also suppress muscle artifacts and motion artifact as shown in Fig. 6.13. In this simulation, the simple implementation of QSW [15], Quad Level Vector (QLV) [16], Continuous Wavelet Transform (CWT) [16], and our previous STAC [14] are modeled using MATLAB. The simulation result shows the proposed method has state-of-the art noise tolerance. and noise database; (a) noise stress test using motion and muscle artifact, and (b) sensitivity (Se) and positive predictivity (+P) of proposed method without noise. The definition of Se is TP/(TP + FN). The definition of the +P is TP/(TP + FP). Then, TP, FN, and FP respectively denote the number of correct QRS complex detection, the number of failures to detect the true QRS complex, and the number of false detection.

6.2.4 Implementation Result

The 3.7×4.3 mm^2 test chip is fabricated using 0.13-um CMOS technology. Figure 6.14 shows the chip micrograph and performance summary. The operating

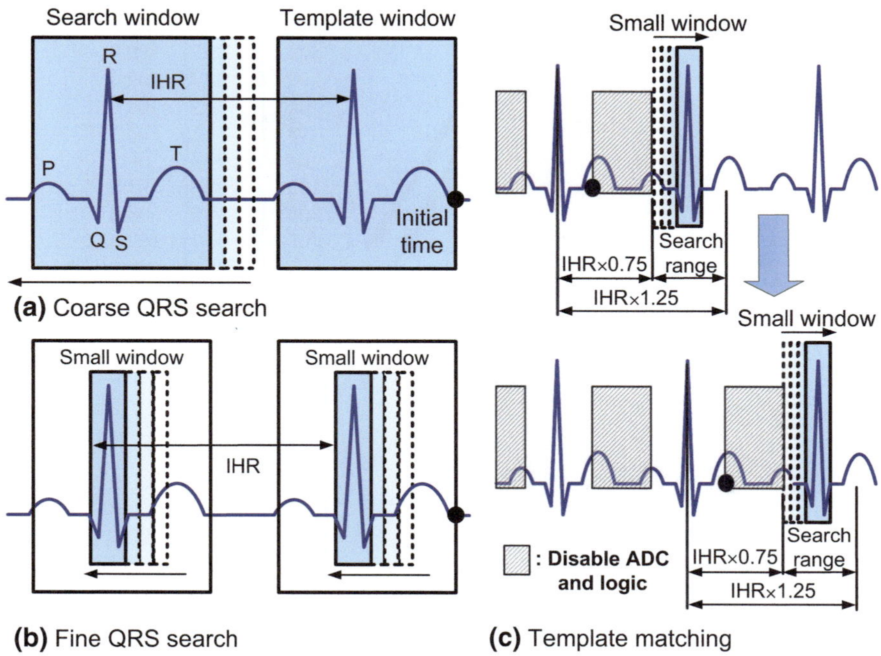

Fig. 6.12 Algorithm overview of coarse-fine QRS template generation and template matching with QRS prediction

voltage is 1.2 V for AFE, ADC, 24-MHz oscillator, IHR extractor, NVMCU, and other digital blocks. Only the 32-kHz oscillator and IO circuits are operated with 3.0-V supply voltage.

To demonstrate the test chip performance, we implemented a heart rate logging application. Figure 6.15 shows the measurement result of IHR extraction waveforms. Then, the sampling rate of ADC is set to 128 Hz and the IHR output is stored to data memory every second. The measurement results show that the IHRs are extracted correctly, even in a noisy condition. Figure 6.16a shows the summary of current consumptions in each block with IHR logging application. The total current consumption is 6.14 uA on average including 1.28-uA nonvolatile MCU and 0.7-uA heart rate extractor. As shown in Fig. 6.16b, the proposed heart rate extractor has higher noise tolerance and minimum power overhead compared with previous noise tolerant hardware [14–16]. Table 6.1 presents a comparison with other recently published ECG monitoring SoCs [15, 17–20]. The proposed SoC has the minimum power consumption in fully integrated (with AFE, ADC, Digital filter, Non-volatile MCU, OSC, and communication I/F) ECG sensors.

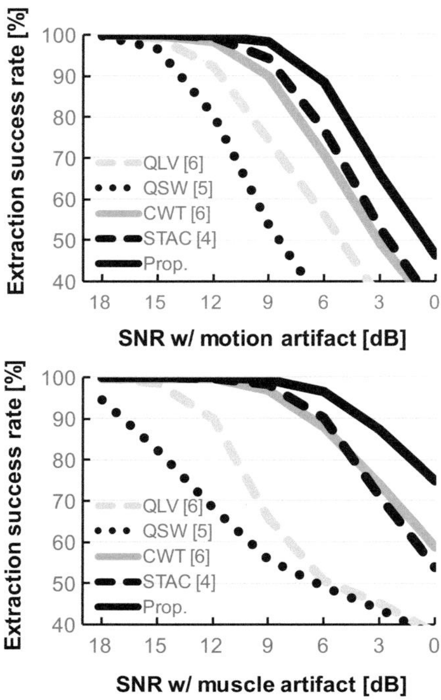

Record	Se [%]	+P [%]
100	99.9	100.0
101	99.7	99.8
102	99.9	100.0
103	99.7	100.0
104	94.4	97.7
105	98.9	97.5
106	95.6	98.6
107	97.7	100.0
108	91.7	90.2
109	99.5	100.0
111	99.9	99.9
112	99.9	100.0
113	99.9	100.0
114	99.6	99.8
115	99.9	100.0
116	99.0	99.6
117	99.9	100.0
118	99.8	100.0
119	99.5	99.8
121	99.7	99.9
122	99.8	100.0
123	99.7	100.0
124	98.0	99.1
avg.	98.8	99.2

(a) Noise stress test **(b) w/o noise**

Fig. 6.13 Success rate evaluation of IHR extractor using MIT-BIH arrhythmia and noise database; **a** noise stress test using motion and muscle artifact, and **b** sensitivity (Se) and positive predictivity (+P) of proposed method without noise. The definition of Se is TP / (TP + FN). The definition of the +P is TP / (TP + FP). Then, TP, FN, and FP respectively denote the number of correct QRS complex detection, the number of failures to detect the true QRS complex, and the number of false detection

6.2.5 Conclusion

We proposed the ECG-SoC using the noise tolerant IHR monitor and NVMCU in 0.13μ m CMOS. The IHR monitor uses the short-term autocorrelation and template matching algorithm for noisy conditions in wearable systems. The NVMCU consists of 16 Kbyte 6T-4C NVRAM and Cortex M0 core with ferroelectric based nonvolatile FFs. The $3.7 \times 4.3\ \text{mm}^2$ ASIC consumes 6.14 μA for the IHR extraction and logging application. The proposed IHR extractor achieves state-of-the-art noise tolerance and power consumption. This result mitigates the performance requirement for analog front end and electrodes. Although other similar works do not consider the noise contamination in wearable and mobile environments, the proposed chip is evaluated in field test under noisy conditions with 5-cm electrode distance and healthy (but not at rest) subject. The wearable system using the proposed 6.14 μA SoC can realize one-week continuous sensing only using a 1-mAh thin-type lithium-ion battery.The

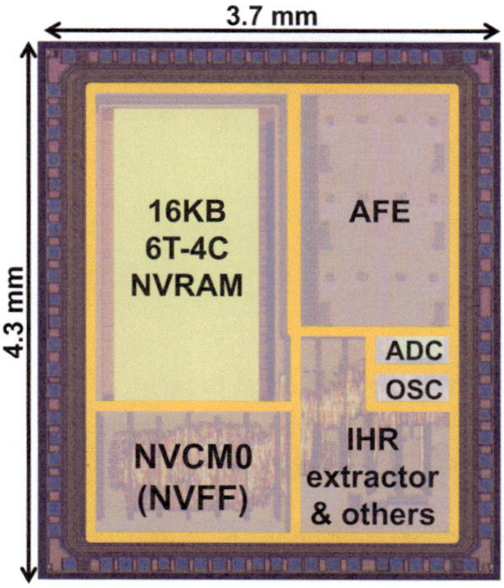

Technology		0.13µm CMOS
Supply voltage		1.2V (AFE, ADC, Logic, Memory)
		3.0V (32kHz OSC and I/O)
Chip area		3.7×4.3mm^2
Frequency		24MHz (for processor)
		32kHz (for other blocks)
Processor		32bit Cortex M0 (with NVFF)
On chip memory		16KB 6T-4C NVRAM
AFE	Gain	54dB
	Bandwidth	700Hz
	CMRR	73dB
ADC	Resolution	8bit
	Current	0.23µA@128S/s, 1.0µA@1kS/s

Fig. 6.14 Chip micrograph and chip specifications

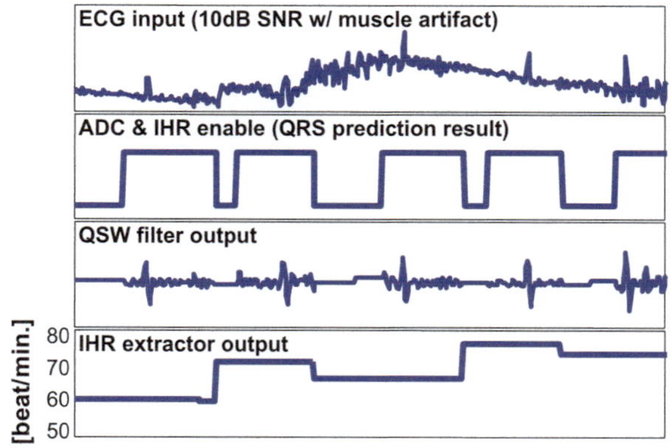

Fig. 6.15 Input ECG waveform and measurement results of QRS prediction, filter output and extracted IHR

processing power can be successfully reduced by proposed normally-off techniques. In the future work, the sensing power should be reduced because it is dominant in the total power consumption.

6.3 Mobile Information Device (MID)

6.3.1 Memory Hierarchy with STT-MRAM for MID

The normally-off computer, an idealistic concept of the ultimate low-power computer with MRAM, was proposed in 2001 [22], in which nonvolatile memory is used for every memory hierarchy, such as flip flops, resistors, resistor files and cache memories

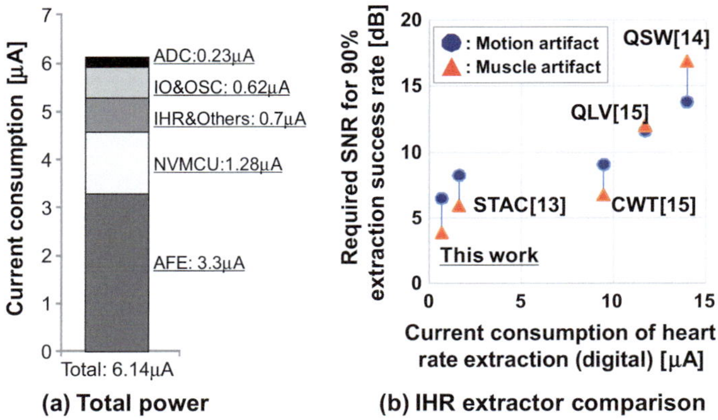

Fig. 6.16 Summary of current consumption; **a** total power consumption and **b** comparison of heart rate extractor performance with previous studies

Table 6.1 Performance comparison with previous studies

	This work	A-SSCC'13 [17]	ESSCIRC'13 [18]	VLSI'13 [15]	ISSCC'13 [19]	ISSCC'12 [20]
Technology	0.13 μm	0.35 μm	0.13 μm	0.13 μm	0.18 μm	0.13 μm
Supply voltage	1.2V/3.0V	2.4-3.0V	1.2V/3.0V	0.5-1.0V	1.8/3.2V	0.3-0.7 V
Frequency	32.768 kHz, 24 MHz	32.768 kHz	32.768 kHz, 24 MHz	8-32 kHz, 24/40 MHz	20 kHz	1.7 MHz-2 kHz
MCU	32b NVCM0	n/a	32b CM0	32b Andes N9	n/a	8b RISC
On-chip memory	16 kB NVRAM	n/a	129.75 kB	20.5 kB	n/a	5.5 kB
Total power for 1-ch IHR	8.47 μW	9.6 – 12 μW	18.24 μW	16.1 μW	18.7 μW	19 μW
Total current for 1-ch IHR	6.14 μA	4.0 μA	13.7 μA	> 16.1 μA	10.41 μA	> 27 μA

in computer systems and power is always shut down whenever CPU core is in an idle state. When the processor is in an idle state, logic data are stored in these MTJ-based nonvolatile circuits, and then power supply is cut off to the logic circuits to stop leakage current. However, since both energy and time for storing data into MTJs in MRAM or STT-MRAM are much larger than in the case of conventional SRAM, when the clock frequency is high as for advanced mobile information devices (~GHz), total power of nonvolatile circuits is not decreased but rather increased compared with that of volatile circuits. For upper-level memory hierarchy, since active power is dominant, it is hard to reduce the power using nonvolatile circuits.

Also, the processor performance deteriorates. Thus, considering the overhead due to write latency and write energy of even the most advanced STT-MRAM, SRAM or STT-MRAM has to be selected for each level of cache memory.

For conventional processor architecture, volatile memories based on SRAM are used for high-frequency access memories such as resistor files, L1 cache memory, and L2 cache memory. Based on the conventional PG technique, power supply is cut off to CPU cores but power supply is not cut off to L2 cache to keep data in it (transition to "CPU core sleep state") during relatively short standby state, where the average power is x2 to x3 lower than that in active state. When time of standby state is long, power supply is also cut off to L2 cache except for small SRAM blocks to retain resistor state. This is a critical transition to "deep power down state (DPS)" having more than x10 lower power than that in active state. Clearly, frequent state transition to DPS plays a vital role in reducing processor power effectively. However, since L2 cache cannot retain data during DPS, after "wake-up" the processor system must recover the lost data from main memory. Consequently, the "wake-up time", time needed for recovery from DPS, is long, which can degrade performance greatly. Therefore, there is a severe tradeoff between power and performance based on conventional PG. Although speed and frequency for PG have been improved, ratio of leakage power for L2 cache to total consumed power for CPU increases with increasing speed of PG, since it is fundamentally difficult to perform PG for volatile cache. It is, hence, considered that L2 or L3 cache memory capacity strongly affects the average mobile processor power determining battery life.

Cache memory capacity has been increased as shown in Fig. 6.17. This can be attributed to two processor architecture trends: Firstly, since it is currently difficult to increase clock frequency over 2–3 GHz from the limitation due to power consumption and heating for the processor, cache memory capacity has been increased to improve the performance. Secondly, for multi-core processors based on conventional architectures, shared cache memory capacity, L2 or L3, is increased with increasing the

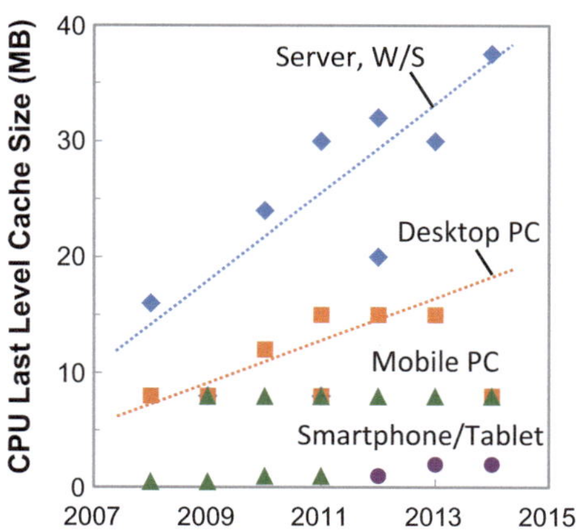

Fig. 6.17 Trend of increasing memory capacity in last level cache

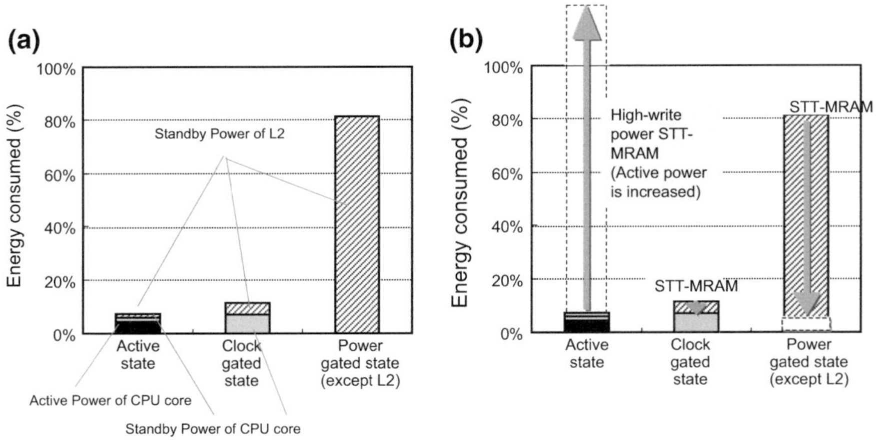

Fig. 6.18 Consumed energy analysis of each state in a conventional low power mobile processor **a** and its change by applying STT-MRAM to nonvolatile L2 cache **b**

number of cores. As a result, power consumed by cache memory occupies the major part in recent processor chips.

Considering such situation of mobile processors, the STT-MRAM/SRAM mixed memory hierarchy has been proposed [21, 23], where SRAM-based cache memory is applied to level-1 (L1) cache, and STT-MRAM-based cache is applied to level-2 (L2) cache, L3 or last-level cache. This is because, whereas active power is dominant for L1 cache because of frequent access to L1, L2 cache operates infrequently, only in the case of L1 cache miss. Further, L2 cache capacity is more than 10 times that of L1. Leakage power in all cache memory is, therefore, dominated by L2 cache (Fig. 6.18 a). To reduce the leakage power in CPU, STT-MRAM should be applied to L2, L3 or last-level cache and write-power of STT-MRAM has to be reduced to decrease the cache active power (Fig. 6.18 b).

This selection varies among applications. For this selection, break-even time (BET) calculated for each application (operation frequency) [24] is hence useful. Figure 6.19 shows an example for a case study of L2 cache. In this case study, three kinds of SRAM having different performance with different order of leakage current are used. BET is changed by cache level, cache capacity and clock frequency. When BET is over the average access interval time of the L2 cache, STT-MRAM should be selected for consumed energy reduction of cache memory.

6.3.2 *How to Reduce Power for Each Low-Power Processor State with STT-MRAM-Based Cache*

Based on BET analysis, memory hierarchy can be selected with SRAM/STT-MRAM. However, advanced processors have multiple low-power states depending on their

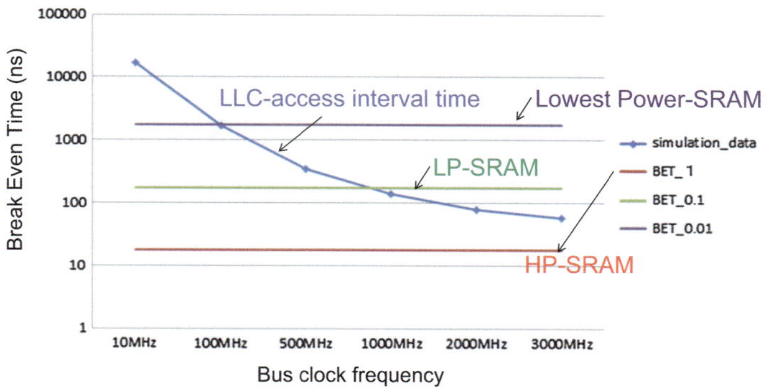

Fig. 6.19 Breakeven time (BET) in comparison of total consumed energy between STT-MRAM and three kinds of SRAM, and simulated LLC-access interval time (LLC-AIT). When BET<LLC-AIT, total consumed energy can be reduced by using STT-MRAM instead of SRAM

idle times. This section describes how to effectively reduce power for each low-power processor state with STT-MRAM-based cache. Figure 6.20 shows conventional low-power states with power-gating that depends on idle times, where the processor has a counter for evaluating the idle time and shifts to lower power state one by one when idle time is longer than a threshold time. Here, C0 is the active state, C1 is clock gated state. C4 is a CPU-core power-off state, where the cache memory outside the CPU-core is power on. For a longer idle time, the processer shifts to partly power-off state for the cache with cache decay mode and finally to entire power-off state for the cache in C6 state, deep power down state (DPS). In C6 state, state SRAM is only active to save CPU state for restarting CPU operation after CPU-core power-on. For deeper sleep state, recovery time and power are increased.

For a processor with STT-MRAM-based cache memory, it should be noted that power of each state can be decreased with STT-MRAM, since standby power of SRAM can be saved for every state, even for active state (C0). Clearly, for sleep states after C4, processor power (standby power) is close to zero. The most effective state for the STT-MRAM to decrease processor power is, therefore, C4 state, as shown in Fig. 6.20.

The "wake-up time", time needed for recovery from the DPS, is long, which can degrade performance greatly. When nonvolatile (NV-) cache is applied, wake-up time from DPS can be decreased to less than 100 ns based on our analysis, as cache data can be retained in nonvolatile memory without power supply. It can increase frequency for power gating, since frequency of standby state is increased with a decrease in standby state time. This fact enables power to be further reduced, since power down is feasible even for short standby state of less than 100 ns.

Processor power has been compared between SRAM-based cache and STT-MRAM-based cache according to the low-power-state models shown in Fig. 6.20. Figure 6.21 shows simulation results for three cases of total processor energy consumption according to the low-power state transition shown in Fig. 6.20. [25] These

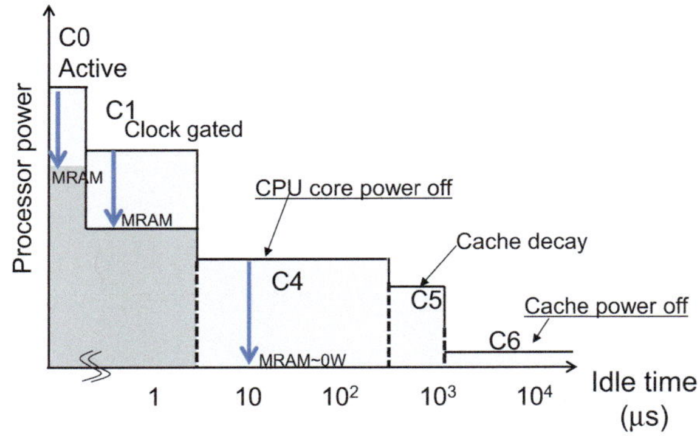

Fig. 6.20 Low-power states with power-gating and STT-MRAM-cache vs. CPU idle time

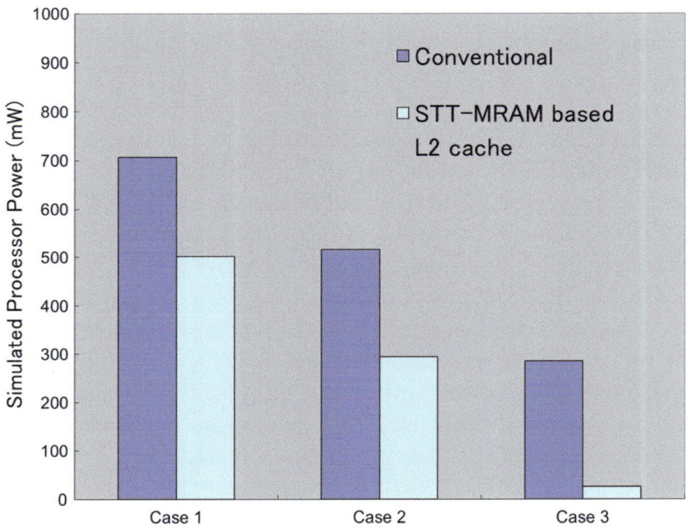

Fig. 6.21 Simulation results on three cases of total processor energy consumption. CPU active time ratios are 52, 28, and 2.5% for case 1, 2 and 3, respectively

three cases were picked up from use case samples for application use test models. CPU active time ratios are 52, 28, and 2.5% for case 1, 2 and 3, respectively. Case 1 is the CPU active state dominated, case 3 is the CPU idle state dominated, and case 2 is midway between these two cases. The result of case 1 indicates that the average power is reduced by 29%, which is attributed to the reduction of leakage power in memory core in L2 cache. It should be noted that normally-off memory core design can, thus, reduce even active power in CPU. In case 2, ultra-fast PG enables the processor to shift to DDS quickly with the NV-L2 cache, whereas the state for the conventional one remains in active state or clock gated state. As a result, the average power is reduced by 43% using NV-L2 cache. In case 3, whereas CPU state data is retained in NV-cache without power supply during long standby state,

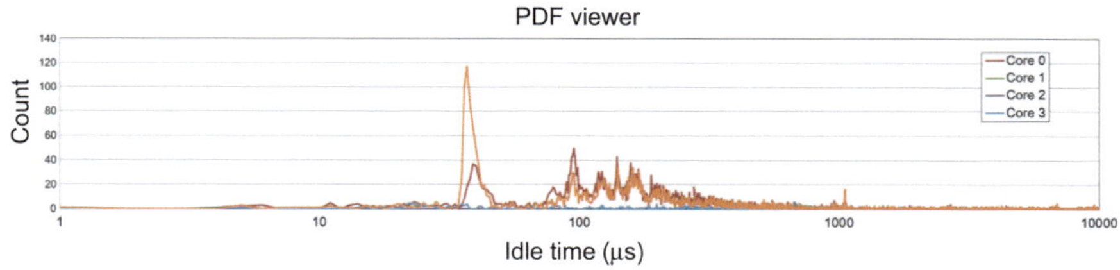

(a) PDF file viewer

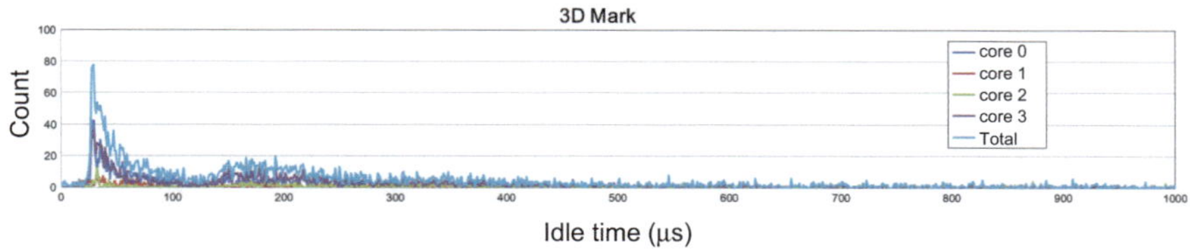

(b) 3D graphics playback

Fig. 6.22 Measured histogram of CPU core idle time with a mobile application processor with quad CPU cores **a** PDF file viewer **b** 3D graphics playback

small-capacitance SRAM with power supply has to be used during it. Consequently, the average power is reduced by 91% using NV-L2 cache. These case studies indicate that STT-MRAM-based cache is a highly effective method to reduce processor power in every situation.

Figure 6.22 shows measured histograms of CPU core idle time with a test board for an advanced mobile application processor [26] while mobile device software, a PDF file viewer (a) and a 3D graphics playback (b), was used. It indicates that major idle times are between 10 us and 500 us. Since these idle times are assigned to C4 state, processor power is expected to be effectively reduced with STT-MRAM cache. This measurement also indicated that 83 and 36% of total time is idle time for software (a) and (b), respectively. Such large ratio of idle time is a typical characteristic of mobile application processors. To estimate the power reduction, simple power gating models were used as shown in Fig. 6.20. As a result, 88 and 67% power reductions were expected for the software (a) and (b), respectively, by replacing SRAM with STT-MRAM for last-level cache. Various other kinds of software for mobile devices were also analyzed. Figure 6.23 shows the relationship between CPU idle time ratio (=idle time/total time) and power reduction rate. It suggests that longer idle time is more effective for power reduction with the STT-MRAM cache scheme. Thus, mobile information devices are among the most effective applications for embedded STT-MRAM memory solutions based on the normally-off policy. However, it has been reported that other computing systems such as high-performance servers, desktop PCs, IoT devices and wearable devices are also candidate applications for STT-MRAM based on normally-off computing [27, 28].

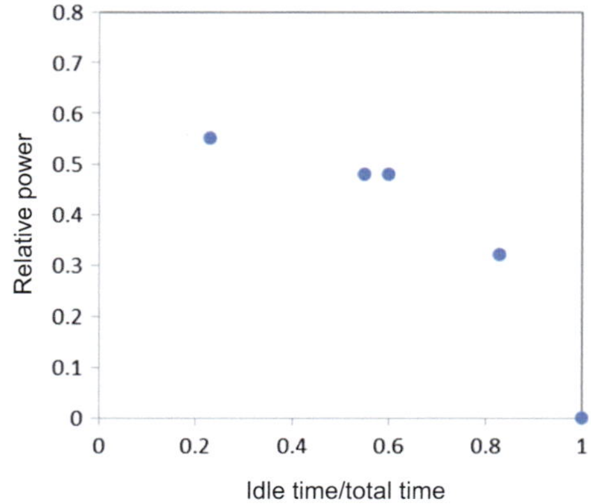

Fig. 6.23 Relationship between CPU idle time ratio (=idle time/total time) and power reduction

6.4 Sensor Node for Social Infrastructure

6.4.1 Sensor Network System for Smart City Application

In this section, we show the feasibility of reducing power consumption by normally-office computer computing control toward the low power consumption of sensor terminal, was taking advantage of environmental data acquisition of the sensor, as an application of these sensors such terminals I will explain about the possibility of low-power consumption at the system level.

A. Normally-off architecture for low-power sensor node

Normally-off architecture for future low-power sensor node is consists of followings and shown in Fig. 6.24.

- normally-off power manager (a),
- sensors,
- sensor controller (b),
- sensor data buffer (c),
- microcontroller.

In this work, normally-off power manager (a) and sensor controller (b) and sensor data buffer (c) are applied to the conventional sensor node. And the roles and an example of the processing of these elements are described.

(a) Normally-off Power Manager
Normally-off power manager manages the task scheduling and power control of sensor, sensor controller, sensor data buffer and microcontroller.
(b) Sensor controller
The relatively simple processing of the data sampling, etc., which were carried out in the microcontroller so far, are performed in the sensor controller.

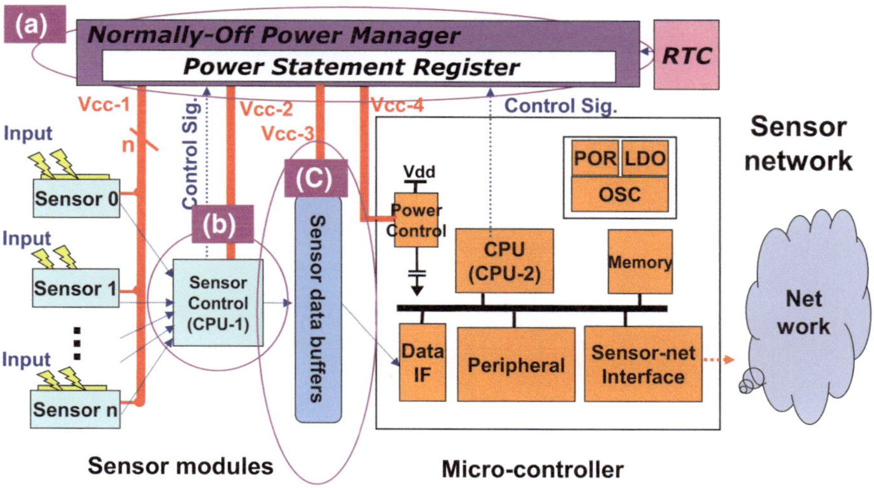

Fig. 6.24 Normally-off architecture for low-power sensor node

(c) Sensor data buffer
The data processed in sensor controller (b) are temporarily stored in sensor data buffer, and are processed in microcontroller only when needed for the averaging process, etc. Therefore, it is possible to reduce the operating frequency of the microcontroller, and to maximize the power-off time of microcontroller.

In the normally-off architecture in this work, the co-design of hardware technology (optimization of power control) and software technology (maximization of power-off time by the task scheduling) is important in order to reduce power consumption effectively in sensor node.

B. Normally-off power management scheme

The target of normally-off power management in this work is to achieve optimal power control in consideration of the breakeven time (BET) in sensor nodes and new ideas of two are introduced. One is an autonomous standby mode transition technology, which proposed in Sect. 5.4, and another is a task scheduling technology, which proposed in Sect. 5.5. By these ideas, it becomes possible of energy optimization in a multi-sensor network system and of usability improvement by ease of application software development.

The activity localization technology by using task scheduling method is described to be utilized for the low-power sensor application.

In the conventional power gating (PG) control, the microcontroller is activated when every data sampling process of sensors is performed (Fig. 6.25). Therefore, total of power consumption energy is large with increase of number of power-on/off cycles of microcontroller.

In this proposed hierarchical power gating (PG) control, sensing data is buffered in sensor data buffer once, and after then microcontroller is activated and performed

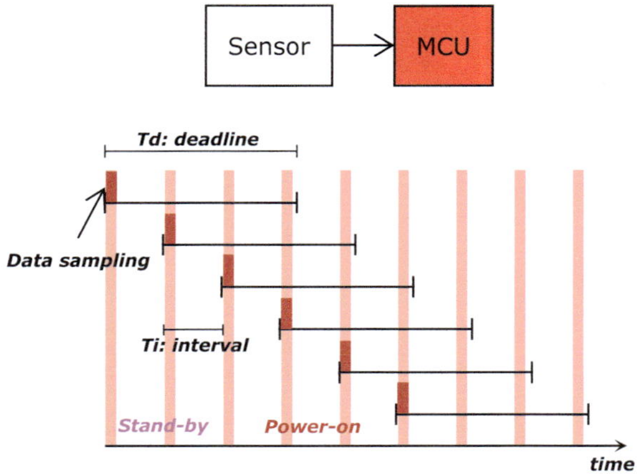

Fig. 6.25 Conventional power-gating (PG) control

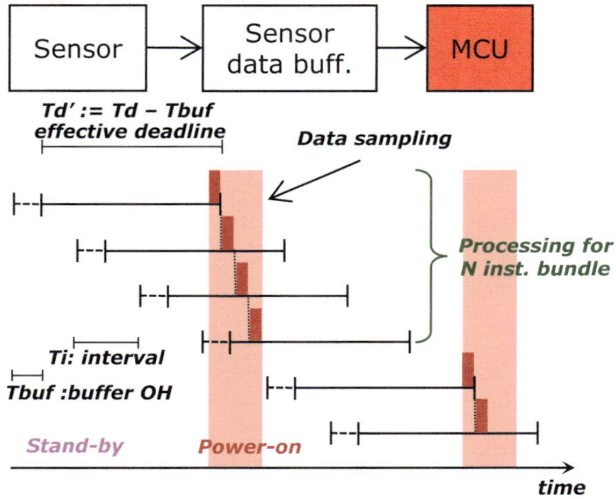

Fig. 6.26 Hierarchical power-gating (PG) control

the process together (Fig. 6.26). Therefore, it is possible to optimize the number of power on/off cycles and much decrease total of power consumption energy.

In our proposed normally-off architecture for low-power sensor node, the activity localization can be realized with effects of sensor controller **b** and sensor data buffer **c**.

C. Evaluation Results

We evaluated the low-power effects at sensor application in our proposed normally-off architecture by using the evaluation board (Fig. 6.27).

The evaluation board is consist of followings.

- Sensor (Temperature, Brightness, Pyroelectric(Motion),
- Microcontroller,
- Other peripheral circuits.

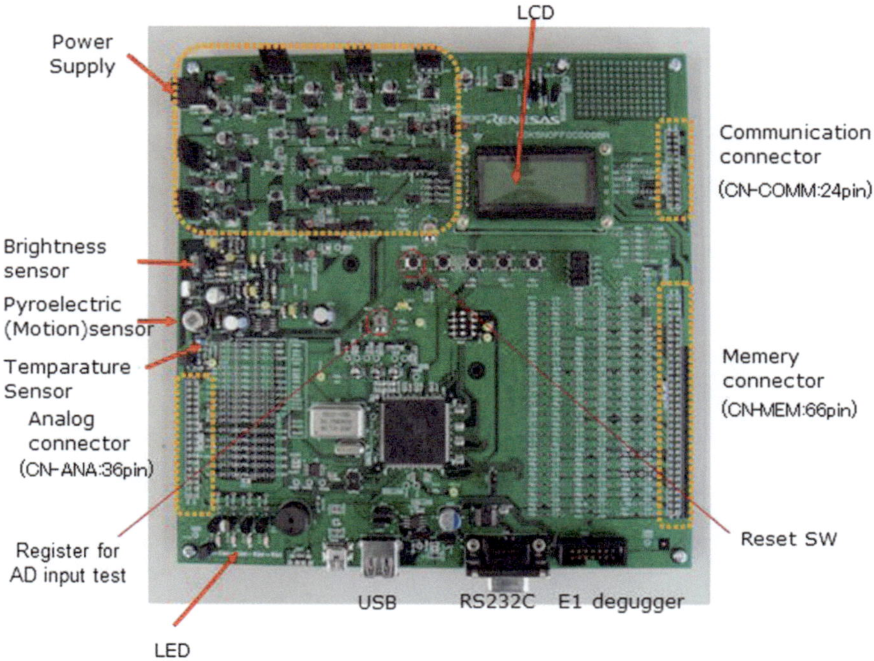

Fig. 6.27 Normally-off evaluation board

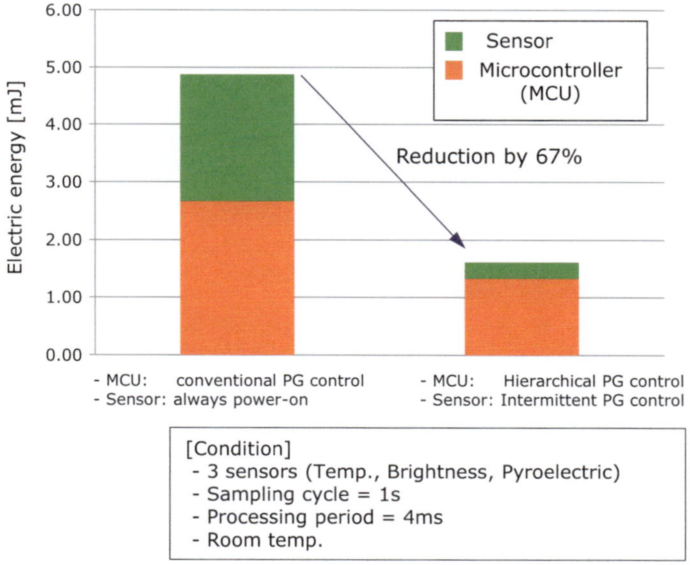

Fig. 6.28 Evaluation results

Evaluation results are below (Fig. 6.28). In conventional sensor node, sensors are always power on, and microcontroller (MCU) is applied with conventional power gating (PG) control. In this work, sensors are intermittent power on, and MCU is applied with the proposed hierarchical PG control.

As the results, it can be obtained the power consumption energy reduction of 67%.

The lower consumption of active currents is possible by reducing the operating time of the microcomputer further, and a further power reduction in sensor nodes is expected by task optimization in normally-off architecture.

References

1. Normally-off computing project (in Japanese). http://www.nedo.go.jp/activities/ZZJP_100016.html
2. Nakajima, H., Shiga, T., Hara, Y.: Systems health care. In: Proceedings of IEEE SMC, pp. 1167–1172, Oct. 2011
3. Oshima, Y., Kawaguchi, K., Tanaka, S., Ohkawara, K., Hikihara, Y., Ishikawa-Tanaka, K., Tabata, I.: Classifying household and locomotive activities using a triaxial accelerometer. Gait Posture **31**, 370–374 (2010)
4. Roel, W., John, M.: Comparing spectra of a series of point events particularly for heart rate variability data. IEEE T-BME, BME-31 **4**, 384–387 (1984)
5. Yazaki, S., Matsunaga, T.: Evaluation of activity level of daily life based on heart rate and acceleration. In: Proceedings of SICE, pp. 1002–1005, Aug. 2010
6. Itao, K., Umeda, T., Lopez, G., Kinjo, M.: Human recorder system development for sensing the autonomic nervous system. In: Proceedings of IEEE Sensors, pp. 423–426, Oct. 2008
7. Itao, K., Ito, T.: Integrated sensing systems for health and safety. In: Proceedings of DTIP Symposium, pp. 212–216, May 2011
8. Zhang, F., Zhang, Y., Silver, J., et al.: A Battery-less 19 uW MICS/ISM-band energy harvesting body area sensor node SoC. In: ISSCC, pp. 298–299, Feb. 2012
9. Kim, H., Yazicioglu, R.F., Kim, S., et al.: A configurable and low-power mixed signal SoC for portable ECG monitoring applications. In: VLSI Symposium, pp. 142–143, Jun. 2011
10. Hsu, S.Y., Chen, Y.L., Chang, P.Y., et al.: A Micropower biomedical signal processor for mobile healthcare applications. In: Proceedings of IEEE ASSCC, pp. 301–304, Nov. 2011
11. Kimura, H., Fuchikami, T., Marumoto, K., Fujimori, Y., Izumi, S., Kawaguchi, H., Yoshimoto, M.: A 2.4 pJ ferroelectric-based non-volatile flip-flop with 10-year data retention capability. In: Proceedings of IEEE A-SSCC, Nov. 2014
12. Masui, S., Yokozeki, W., Oura, M., Ninomiya, T., Mukaida, K., Takayama, Y., Teramoto, T.: Design and applications of ferroelectric nonvolatile SRAM and Flip-FF with unlimited read, program cycles and stable recall. In: Proceedings of IEEE CICC, pp. 403–406 (2003)
13. Hsu, S.Y., Chen, Y.L., Chang, P.Y., Yu, J.Y., Yang, T.F., Chen, R.J., Lee, C.Y.: A micropower biomedical signal processor for mobile healthcare applications. In: Proceedings of IEEE A-SSCC, pp. 301–304, Nov. 2011
14. Fujii, T., Nakano, M., Yamashita, K., Konishi, T., Izumi, S., Kawaguchi, H., Yoshimoto, M.: Noise tolerant instantaneous heart rate and R-peak detection using short-term autocorrelation for wearable healthcare systems. In: Proceedings of IEEE EMBC, pp.7330–7333, July 2013
15. Hsu, S.Y., Ho, Y., Chang, P.Y., Hsu, P.Y., Yu, C.Y., Tseng, Y., et al.: A 48.6-to-105.2μW machine-learning assisted cardiac sensor SoC for mobile healthcare monitoring. In: Dig. IEEE Symposium VLSI Circuits, pp. 252–253, Jun. 2013
16. Kim, H., Yazicioglu, R.F., Kim, S., et al.: A configurable and low-power mixed signal SoC for portable ECG monitoring applications. In: Dig. IEEE Symposium VLSI Circuits, pp. 142–143, Jun. 2011

17. Deepu, C.J., Zhang, X., Liew, W.S., Wong, D.L.T., Lian, Y.: An ECG-SoC with 535nW/channel lossless data compression for wearable sensors. In: Proceedings of IEEE A-SSCC, pp. 145–148. Nov. 2013
18. Izumi, S., Yamashita, K., Nakano, M., Konishi, T., Kawaguchi, H., Kimura, H., et al.: A 14 uA ECG processor with robust heart rate monitor for a wearable healthcare system. In: Proceedings of IEEE ESSCIRC, pp. 145–148, Sep. 2013
19. Kim, S., Yan, L., Mitra, S., Osawa, M., Harada, Y., Tamiya, K., et al.: A 20uW intra-cardiac signal-processing IC with 82dB bio-impedance measurement dynamic range and analog feature extraction for ventricular fibrillation detection. In: ISSCC Dig. Technical Papers, pp. 302–303, Feb. 2013
20. Zhang, F., Zhang, Y., Silver, J., Shakhsheer, Y., Nagaraju, M., Klinefelter, A., et al.: A battery-less 19uW MICS/ISM-band energy harvesting body area sensor node SoC. In: ISSCC Dig. Technical Papers, pp. 298–299, Feb. 2012
21. Abe, K., Nomura, K., Ikegawa, S., Kishi, T., Yoda, H., Fujita, S.: Hierarchical nonvolatile memory with perpendicular magnetic tunnel junctions for normally-off computing. In: The 2010 International Conference on Solid State Devices and Materials (SSDM), Tokyo, pp. 1144–1145, Sep. 2010
22. Ando, K.: A Normally-off Computer. FED Journal **12**, 89 (2001). (in Japanese)
23. Ando, K., Ikegawa, S., Abe, K., Fujita, S., Yoda, H.: Normally-Off Computer: new roles of nonvolatile devices in future computer systems. In: Sustainable Green Computing, IGI press, ISBN 978-1-4666-1842-8, June, 2012
24. Takeda, S., et al.: Low-power cache memory with state-of-the-art STT-MRAM for high performance processors. In: ISOCC (2015)
25. Fujita, S., et al.: Novel nonvolatile memory hierarchies to realize normally-off mobile processors. In: 19th Asia and South Pacific Design Automation Conference (ASP-DAC) (2014)
26. Samsung Exynos™ (Exynos 4412) for Galaxy mobiles. http://www.samsung.com/exynos/
27. Fujita, S., et al.: Technology trends and applications of MRAM from big data to wearable devices. In: ISSCC (2015)
28. Fujita, S., et al.: Technology trends and near-future applications of embedded STT-MRAM. In: IMW (2015)

Chapter 7
Related Research & Development

Jun-ichiro Yoshikawa

Abstract Energy consumption is one of the most important problems in the world. To take a general view of activities which focus on energy saving, related research and development in Germany, US and Japan are surveyed and introduced. In each area, different types of activities are observed depending on their policies such as government initiative or private initiative.

Keywords Energy saving project · Research & Development

7.1 Overview

The striking feature of our normally-off computing project is to aim to develop low energy consumption memories and devices using such memory, and also to reduce total energy consumption of whole system. There are many projects in Japan and overseas which target low energy consumption of semiconductor alone including memories, and apparatus in itself, as far as the author checked, no similar project is found. As a result, originality of this project is understood. For reference, the lists of energy saving related national projects in Europe and in USA are listed in [1].

By the way, an industrial-activity becomes active worldwide, which accumulates and analyzes a lot of data. The data is collected from sensors being installed in various devices, infrastructure and so on, and which links the result to the efficient utilization of devices, the maintenance-management of infrastructure, the improvement of user's convenience, etc. These systems are called as *IoT* (Internet of Things). As to the purpose of IoT, it is sometimes pointed out the improvement of the energy efficiencies of the device and the communications infrastructure, and it is also expected that our project result is applied to IoT field. Therefore, in this chapter, we introduce the

J. Yoshikawa (✉)
Yoshikawa Legal Office, Yokohama, Kanagawa, Japan
e-mail: info@yoshikawaj.com

T. Nakada and H. Nakamura (eds.), *Normally-Off Computing*,
DOI 10.1007/978-4-431-56505-5_7

trends to IoT in Germany which is taking a futuristic national-policy, in the U.S. where the individual firm tackles mainly, and in Japan.

7.2 Germany

In Germany, traditionally, the international competitiveness of the manufacturing industry was strong but, relatively, such competiveness declined for the reasons such as the persistently high personnel expenses due to the labor conservation policies and technical power improvement of a developing country in and out of the Europe. The German government which sensed a crisis in this fact, with a coalition of industry-government-academia, deployed various efforts to share the manufacturing data, which is gotten from the production installation at each manufacturing process, with the downstream process and customers in order to reduce the inventory, increase in efficiency of the physical distribution, and improve the quality, as the means of the competitiveness restoration. These efforts are called *Industry4.0* (the 4th Industrial Revolution). This project has a characteristic with the tactical national project which concentrates the total power of industry-government-academia.

For example, at the factory of Bosch, which is the largest motor vehicle part manufacturer in the world, the manufacturing installation reads information on the Bluetooth terminal worn at the waist of the worker and displays a manufacturing manual by the video and characters according to the age and the years of experience. The display language can be changed according to the worker's native language, too. With these efforts, it says that work efficiency improved remarkably, and that it leads to improve competitiveness in response to high-mix low-volume production. Bosch, in the future, aims to reflect an order immediately to the production plan by linking a network to the research and development division, the procurement place and customers in addition to the manufacturing department.

7.3 The U.S.A

In the U.S., similar efforts are not a national policy and are done as management strategy of respective companies. Also, they are characterized by active activities such as making a global standard of interfaces and etc. among the equipment relating to IoT, cooperation aiming at the monopoly and establishing of a consortium.

The main elements are consisting of (i) intelligent equipment, (ii) advanced information analysis, and (iii) worker who executes best operation by drafting it based on the result of analysis, but persistently, the key is the person of (iii) and IT is charged with the supplementary role. On the other hand, IT megacorporation such as Google, Apple and Amazon reviews unique tie-up strategy. The approaches of the main consortium and respective companies are introduced below.

(1) Industrial Internet CONSORTIUM

The Industrial Internet Consortium was founded by 5 companies of GE, IBM, Intel, Cisco Systems and AT&T in May, 2014 by aiming at the utilization of IoT for the industrial field, and is participated by the 157 organizations, as of May, 2015, mainly among U.S. companies.
The purposes of the consortium activity are shown below.

- Drive innovation through the creation of new industry use cases and testbeds for real-world applications;
- Define and develop the reference architecture and frameworks necessary for interoperability;
- Influence the global development standards process for internet and industrial systems;
- Facilitate open forums to share and exchange real-world ideas, practices, lessons, and insights;
- Build confidence around new and innovative approaches to security.

They says that the participating organizations use respective own factory and own infrastructure and the verification tests are proceeded jointly. By the introduction of IoT, it is expected that the improvement of the productivity and functionality enhancements of the machine tool by software. In these organizations, GE is showing active efforts. For example, sensors installed on the aircraft engine lead to prevent an unexpected engine trouble from occurring, by executing as suitable level of maintenance and a repair as required. As a result, the inefficiency of the operation can be improved and flight cancellation rate can be reduced.

(2) ALLSEEN ALLIANCE

It is the standardization organization of the communication among the equipment, which is formed around QUALCOMM Inc. and so on.

(3) OPEN INTERCONNECT

It is the standardization organization of the communication among the equipment, which is led by Intel.

(4) The Thread Group

It is the standardization organization of IoT, which is led by Google.

(5) The trend of Amazon

Amazon doesn't make strategy about IoT clear, but they has the immeasurable capability to collect big data for general consumers and in the industrial field, and it is often reported that the trend of moving into the field of manufacturing in the near future.

(6) The trend of Apple

Apple is proposing, on its own accord, a standard HomeKit to control the iOS-based network equipment for home.

7.4 Japan

In Japan, the independent administrative agency New Energy and Industrial Technology Development Organization ("NEDO") engages it in the form of promotion of the base technology development of Cyber-Physical Systems ("CPS"). CPS can be explained by the system consisting of the physical system and the networked computing.

NEDO, as the information collection business in 2011, carried out an investigation about the consolidation of platform of the technology, the system and the industry in order to promote industrial fusing using ICT (Japanese version of the "Cyber-Physical System".).
The report states as follows.

"In recent years, with the development of the cloud computing and the sensor network technology, the feasibility of "Cyber-Physical Systems" is appealed. This system is to capture information on the real world into the cyber space and to perform advanced analysis and control. Our country is confronting a wide range of compound social problems such as the expansion of medical expenses due to low birthrate and longevity, a decline in Japan's food self-sufficiency ratio and low productivity of service industries, and some says that Japan is the "advanced country with full agenda".

It is expected that the realization of *ICT*-used "Cyber-Physical Systems" creates a new industry in a form of industry-fusing-like and unconventional, and, that it brings about the more intelligent society to contribute for more efficient solutions of social problems. So it is desirable to realize it immediately.

On the other, with the realization of "Cyber-Physical Systems", it is mandatory not only the solution of the technical problems, but also the solution of various kind of institutional problems and moreover the creation of business model which is rooted in the new visions of society.

These problems are not independent each other but are the compound problems getting complexly intertwined with respect to one another, therefore, it is desired certain efforts towards problems arrangement and resolution in the form of unity of the industry-government-academia.

This project, with a background like the above situation, aims to clarify problems for the arrangement of platform which is so concerned about technologies, institutions and industries in order to realize "Cyber-Physical Systems", by assuming specific field in advance.

In this project, the three fields of "agriculture", "medical care" and "service industry" are selected as high-potential expansion field in the future about the "CPS", based on the interim report "Towards the Fusing New Industry" made by the Information Economy Subcommittee of the Industrial Structure Council in the MITI, and on the result of discussions with the knowledgeable persons, further, the six hypothesis of future scenarios are built in connection with these by activities of workshop.

These titles are shown below.

(1) Personal services of medical and healthcare based on biological information and behavior information.
(2) Happiness promotion service based on estimation of mental conditions.
(3) Personal program of eating habits based on health exam.
(4) Safe and pleasant service by sharing the personal information of the family circle.
(5) Revolution of entrance exam and job-hunting? Breakthrough of difficulty is "better personal information".
(6) Building a Smart City based on behavior in daily life by life monitors.

According to the interview with general people about acceptability of these six (6) future scenarios, their highest acceptability was focused on (1) Personal services of medical and healthcare based on biological information and behavior information and (3) Personal program of eating habits based on health exam.

Moreover, the result of this investigation shows, as for these two scenarios, that the majority of general people expect the support/involvement in any form by the national and local governments. Therefore, it is considered that certain rationality lies in investigation of the future national subject, particularly in this field."

In case of Komatsu Machine Tracking System ("KOMTRAX"), which is developed by Komatsu Ltd. ("Komatsu"), is famous as an original activity example by the enterprise. KOMTRAX installs sensors on the construction equipment like the power shovels. KOMATSU, at their head office, is able to monitor the operating conditions of all machines sold to the customers worldwide.

By proactively using this system, Komatsu became able to provide the customers with more efficient maintenance-management services of the construction machines. Also, they can provide appropriate assistance regarding the improvement of fuel consumption of the construction machines to the customers. The customer-satisfaction is also improved by the discovery of stolen construction machines.

Although the individual action such as the above was found, it has only just begun the action to IoT in the cross-systemic approach by framework of industry-government-academia.

Reference

1. NEDO report of "Technical Research toward development and promotion of Normally-Off computing" (in Japanese), Dec 2015, http://www.nedo.go.jp/library/seika/shosai_201512/20150000000747.html

Chapter 8
Conclusion

Takashi Nakada and Hiroshi Nakamura

Abstract In previous chapters, we introduce idea, methodology of "Normally-Off Computing" with a wide variety of examples. In this chapter, we explain expectation and future aspect of normally-off computing.

Keywords Expectation · Future aspect

Low power technologies of computer systems are indispensable for the forthcoming sustainable and sophisticated information society. Normally-off computing is one of the promising ways to achieve this goal. In this book, we described its concept and methodology. In addition, our normally-off computing project is introduced.

We attempt to find the best way to make full use of the new generation non-volatile memories. So far, from the view point of memory system optimization, only two properties of memory are focused on and discussed, that is latency and capacity. Now, non-volatility appears as the third property.

This property absolutely contributes further power reduction. However, it will not effectively lead to power reduction without careful considerations because of two major problems or challenges, that is, memory hierarchy and activity granularity. In other words, once we overcome these problems, further power reduction would be achieved. Cooperation and cooptimization of different design layers, including algorithm, OS, compiler, architecture, circuit and device, are definitely required.

In Chap. 1, we introduced background of "Normally-Off Computing". For future sustainable society, low power technologies are necessary to reduce total energy consumption of IT equipment. To cope with this issue, much attention has been paid to normally-off computing, which aims for extremely low power computing.

T. Nakada (✉) · H. Nakamura
The University of Tokyo, Bunkyo, Tokyo, Japan
e-mail: nakada@hal.ipc.i.u-tokyo.ac.jp

H. Nakamura
e-mail: nakamura@hal.ipc.i.u-tokyo.ac.jp

T. Nakada and H. Nakamura (eds.), *Normally-Off Computing*,
DOI 10.1007/978-4-431-56505-5_8

In Chap. 2, a wide variety of low power techniques are introduced. Any techniques have trade-off between the energy and the performance. When we focus on two low power techniques, their trade-off point is called "Break Even Time (BET)". The BET is one of the most important criteria to realize normally-off computing.

In Chap. 3, several next generation non-volatile memories, which will be used for normally-off computing, are explained. Their properties are summarized and discussed.

In Chap. 4, a design methodology of normally-off computing is explained. Basics of low power design for hardware, architecture and software are discussed.

In Chap. 5, details of low power techniques for each component are introduced. Specifically, techniques for logic, cache, processors, scheduling and data management are introduced. Normally-off computing is realized by combining them.

In Chap. 6, we overview our "Normally-Off Computing Project" supported by NEDO/METI in Japan. Three practical systems with normally-off computing are explained. For healthcare systems, we introduce ultra-low power systems, which designed with non-volatile FFs and aggressive power management. For mobile information devices, non-volatile cache and its power management scheme are explained. For sensor network systems, adaptive power management scheme and task scheduling algorithm are introduced. Through these systems, energy optimization is based on normally-off architecture and significant energy reduction is achieved.

In Chap. 7, we introduce and summarize related projects in the world.

Information society is changing day by day. A lot of researches on new memory devices and low power technologies are performed. Other types of useful devices or technologies other than what introduced in this paper may become available in the future. However, it would not be fruitful if we discuss the device properties only. The viewpoint of "computing technology," that is the technology how to make full use of the attractive property is essentially important. Therefore, our concept and methodology are still useful for future computer systems.

MIX
Papier aus verantwortungsvollen Quellen
Paper from responsible sources
FSC® C105338

If you have any concerns about our products,
you can contact us on
ProductSafety@springernature.com

In case Publisher is established outside the EU,
the EU authorized representative is:
Springer Nature Customer Service Center GmbH
Europaplatz 3, 69115 Heidelberg, Germany

Printed by Libri Plureos GmbH
in Hamburg, Germany